ALL THESE WORLDS ARE YOURS

JON WILLIS

All These Worlds Are Yours

The Scientific Search for Alien Life

Yale UNIVERSITY PRESS

NEW HAVEN AND LONDON

Yale University Press books may be purchased in quantity
for educational, business, or promotional use. For
information, please e-mail sales.press@yale.edu (U.S.
office) or sales@yaleup.co.uk (U.K. office).

Set in Janson and Futura Bold type by Westchester
Publishing Services.
Printed in the United States of America.

Library of Congress Control Number: 2015959236
ISBN 978-0-300-20869-6 (hardcover : alk. paper)

A catalogue record for this book is available from the
British Library.

This paper meets the requirements of ANSI/NISO
Z39.48-1992 (Permanence of Paper).

10 9 8 7 6 5 4 3 2 1

To Rose and Sara.
Thank heaven for little girls.

CONTENTS

PREFACE

All these worlds are yours, except Europa. Attempt no landing
there. Use them together. Use them in peace.

All these worlds are yours. The idea behind the title is that the
search for life in the universe, both its practicalities and its implications, is accessible to anyone with an interest in science. The quote
is taken from Arthur C. Clarke's *2010: Odyssey Two*, and the use of a
science fiction novel is deliberate.

Many of our preconceptions regarding alien life have been formed
within the imaginations of writers and filmmakers. Such acts of creative imagination are often quite stunning and profound, but perhaps in ways that reveal the nature of the human creator rather than
some unknown alien. One result, however, is that we all have some
idea of what forms alien life might take, whether they will greet us
or just eat us, or perhaps greet us then eat us.

Such ideas may well be anthropocentric, but when teaching the
search for life in the universe to university students, I am always
impressed by how eager the students are to move beyond popular
imagination and how excited they are to realize that science fact
outstrips science fiction in coming up with new and radical ideas.

Hence this book.

ACKNOWLEDGMENTS

Sincere thanks are due to Joe Calamia and all of his colleagues at Yale University Press for bringing this book to you. Many others have provided much-needed help along the way, and I wish to thank them now.

Thank you to Florin Diacu and Paul Zehr at the University of Victoria for their invaluable advice and encouragement, which saw this book progress from an idea to a formal proposal.

Thank you to J. J. Kavelaars, Dave Patton, James DiFrancesco, and Christian Marois for supporting my book proposal and writing their encouragement to prospective publishers.

While many people have enriched this book through many discussions, I want to give particular thanks to Colin Goldblatt for his unceasing enthusiasm and ideas and for contributing much that has taken root in this book.

Many thanks to those who have read this book in part or full and been very generous with their comments and encouragements: Jillian Scudder, Kim Venn, Maggie Lieu, Trystyn Berg, Chelsea Spengler, Jericho O'Connell, Jeremy Tatum, Colin Scarf, Sebastien Lavoie, Kyle Oman, and Michele Bannister.

Finally, many thanks to James Kasting and one anonymous reviewer who conducted formal reviews of this book and provided valuable and detailed feedback.

ALL THESE WORLDS ARE YOURS

Do aliens exist? Is there life beyond planet Earth? Well, yes, there is. Probably lots of it. How can I be so sure? Sure enough to start the book by telling you the answer? My answer is in large part based on a mathematical argument. The universe, as we shall see, is a very large place—quite possibly infinite in size. We don't need to go into the math to understand that infinity means big. Big enough that, even though the possibility of something occurring—such as life—may be almost impossibly small, it must occur somewhere. The odds on your lucky lottery number may well be vanishingly small, but as long as the chance is not zero, if you play an infinite number of times, you are guaranteed to win. In an infinite universe, everything is possible. In many ways, though, this is an unsatisfying answer—one that locates alien life in the distant recesses of what we imagine the cosmos to be. Much more interesting are the questions where is alien life to be found, what forms does it take, how does it live (and breathe), and how should we interact with it? But as we shall discover throughout this book, the answers to these questions are much more challenging to obtain than my rather facile yes in response to the question, do aliens exist?

What happens if I change the question slightly? Is there any scientific evidence that life exists anywhere in the universe beyond Earth? The answer to this question—at present—is emphatically no. It is possible that this is because life does not exist anywhere in the universe except Earth. But given my certainty expressed above, it is more likely that life indeed exists elsewhere in the universe, but we are yet to discover it. We haven't yet searched, scoped, poked, or peered into enough places in the universe to have found it. For full disclosure, I should note that it is also possible that we already have the scientific evidence of life beyond Earth but that such evidence is not universally accepted as being proof positive—more later.

Even once you have finished reading this book, the answer to my second question will probably still be no.[1] In large part this is because the challenge is great compared to our current resources. Despite the claims of UFO enthusiasts, alien life has not spontaneously appeared on our doorstep. Alien life also appears to be beyond the current reach of our telescopes and space probes. In a world of limited scientific resources, we have to decide exactly where to search and how to search in order to achieve the best chance of success. Scientists refer to the ideas that underpin these efforts as astrobiology. The science of astrobiology has three main goals: to understand the conditions necessary for life on Earth (and possibly the conditions required by life in general), to look for locations in the universe which supply these conditions, and, finally, to detect life in these locations. At present we have discovered a great many potential habitats for life, planets and moons in our solar system or planets orbiting distant stars. Some of these new worlds replicate in part conditions we encounter on Earth—the only location in the universe where we know that life exists.

At this point, the keen reader will protest as to why we use life on Earth as a template for the search for life in the universe. What if life on Earth represents only a tiny fraction of the range of properties displayed by life beyond Earth? Is our search too narrow in scope? Do we run the risk of missing truly alien life because we don't know how to recognize it? Once again, the answer is clearly yes. By starting with life on Earth and working outward, our search will not discover all possible life-forms. We will overlook the consciousness of free floating asteroid-like organisms and other unthought-of possibilities. But you have to start somewhere. The only life we know is that of Earth. By using this as a starting point, we can speculate about the kinds of life processes that might occur on planets similar to Earth—and by this I mean those that have a solid surface, an atmosphere of some kind, and possibly a range of chemicals in the liquid state. The only thing I can guarantee is that the more we look, the more we will learn about life and its possibilities. Okay—protest postponed? Let's proceed.

We therefore stand on the threshold of asking whether life exists in specific locations in the universe—with all the attendant challenges of such an important yet basic question: What physical tests are required to unambiguously confirm the existence of what may well be an unfamiliar life-form? What technology is required to perform such an investigation? Do we need to get physically up close and personal with the organisms in question, or can we learn about them remotely? Questions, questions, questions. The answers to which draw on a fascinating range of subjects: astronomy, physics, chemistry, biology, geology, mathematics, computer science, and philosophy, to name but a few. Although the range of concepts employed by astrobiologists is indeed broad, the ideas they represent remain very accessible. Though the ideas are certainly not what I would call simple, they do lie within the grasp of any scientifically literate person. So with that in mind, the idea I want to start with is a very simple yet very old one. It is the idea of a world.

New Worlds for Old

A world is place. You can experience it, explore it, pick it up, and shake it. In some sense it is the "real" world as opposed to an abstract idea. For much of human history, Earth was the only such world. Hence, we explored it, discovering new cultures and life-forms. The stars and planets visible to the naked eye of classical astronomers represented a noble field of study.[2] But the stars and planets they observed were ultimately considered points of light within a dark firmament. Myth and speculation attempted to describe their nature, but such ideas remained abstract, unobserved, and consequently imaginary. What changed was vision. We looked closer and saw more. And the more we saw, the more we realized that each planet and each star was a real place, governed by physical processes identical to those that had shaped Earth and our sun. We could therefore encounter, investigate, and explore such worlds. We could visit them and stand on their surfaces. We could meet their inhabitants.

Galileo Galilei was the first person to provide that vision. In 1609, Galileo tried to convince the merchants of Venice that his apparatus of two lenses attached to either end of a wooden tube was a good way to observe ships at a distance. While you and I would call the device a telescope, Galileo marketed the idea under the catchy name perspicillum. From his vantage point atop the bell tower of St. Mark's, he could observe ships approaching Venice up to a day's sailing from port. The magnification offered by his telescope allowed him to identify each approaching ship by observing its flags and pennants, information he offered to sell to their owners, who thus would have a day's head start on their competitors. It is not clear how much success he had in persuading the merchants of Venice to get a jump on the market, but at some point he decided to do something very different and very interesting. He pointed his telescope at objects in the night sky—among them the moon, and the planets Jupiter and Saturn. Second, and critically for everyone thereafter, in 1610 he described what he saw in the simple, unadorned prose of the *Sidereus Nuncius*, or *Starry Messenger*.

The moon turned out to be a world, a rather interesting one with craters, jagged, unruly mountains, and deep, shadow-filled valleys. He suggested that the smooth, featureless plains were seas, *maria* in the Latin of his text, the name we still use today. Importantly, all this was based on direct observation with his telescope. Until this point, the moon had been largely unknown except for its regular motion about Earth. Classical astronomers had considered it a perfect, unblemished piece of creation, appropriately for an inhabitant of the celestial sphere. Instead, it turned out to be pockmarked, jagged, eroded, and resurfaced—much as Earth was. Imperfect, complicated, and therefore interesting!

From Galileo's first observations of the moon, it took us only 360 years to get around to visiting the place—poking, prodding, and (what will turn out to be of great importance for our exploration of the universe) bringing bits of it back to Earth. From our numerous manned and robotic explorations, we have learned that the moon is made of rocks very similar in composition to the outer crust of Earth. As far as we can tell, the oldest lunar rocks are the same age as the oldest rocks on Earth—about 4.4 billion years old—and slightly younger than the meteorites that appear to be the oldest chunks of debris in the solar system (about 4.54 billion years old). The inference based on these observations is that Earth and the moon were, at some early point in their history, one lump of molten rock. Some event, perhaps a collision with another planetoid in the young solar system, broke the Earth-moon planet apart, the larger and smaller fragments becoming Earth and the moon, respectively.

In the 360 years bookended by Galileo Galilei and Neil Armstrong, the moon went from being a familiar yet poorly known voyager through the heavens to being a solid body with a long geological history ultimately twinned with that of Earth. Long before man stepped foot on the moon, it had entered our consciousness as a world—a physical part of our experience of nature. Distant, yet real and tangible.

The Chances of Anything Coming from Mars . . .

Are a million to one. So pronounces the astronomer Ogilvy in H. G. Wells's *The War of the Worlds*. Wells published the novel in 1896 for a public primed to consider Mars the next world in our growing collective consciousness of the universe beyond Earth. The man who more than any other prepared the public to believe in an alien civilization on Mars was Percival Lowell. His story is important for the search for life in the universe, although, as we shall see, his claims taught us that one should also bear in mind at which end of the telescope that life is.

Percival Lowell eludes the simple sketch of biographical caricature. Though his ideas regarding life on Mars were ultimately flawed, he remained a serious scholar. Lowell could be labeled an amateur astronomer who formed part of a long tradition of independently wealthy men whose resources, coupled with their scientific drive, propelled many to become leading contributors to their field of interest. His decision to site Lowell Observatory, with its suite of telescopes, in the dark skies of Arizona rather than close to the practical amenities of a large city[3] looked forward to the modern, professional age, when telescopes are sited where their scientific observations can be most effective.

Lowell's interest in astronomy was directed toward the planet Mars—a passion fired by the work of his peer Giovanni Schiaparelli. Schiaparelli, the director of the Milan observatory, began his observations of Mars with the "great opposition" of 1877. In astronomy, opposition occurs when Earth and Mars (for example) line up on the same side of the sun. The two bodies are often at their closest at this point, and thus opposition makes for good planetary viewing.

When viewing Mars through a refracting telescope of the aperture used by Schiaparelli and Lowell,[4] the planet appears as a pale pink disk—with dark blemishes apparent if the volcanic Tharsis plateau is in view. Depending on the point in the Martian year in which one observes it, Mars will display the bright white of a polar ice cap that ebbs and flows as the seasons change. In addition, Mars experiences planetwide dust storms that periodically obscure the entire surface, leaving what appears to be a featureless disk. Schiaparelli claimed that when observing Mars, he was able to resolve dark linear features on the surface, which he referred to as *canali* (channels). Schiaparelli noted that these features would resolve themselves in his telescope eyepiece only at brief moments of apparent atmospheric calm, when the blurred, restless image of Mars, agitated by waves of air pressure in Earth's atmosphere, stilled itself and revealed what appeared to be a clear view of the surface.

So far, so scientific. Schiaparelli observed and reported what he saw faithfully. Though he speculated about the nature of these *canali*, he was balanced and cautious. But Schiaparelli's observations become the point from which Lowell would make a fateful leap of speculation. He claimed that Schiaparelli's *canali* were real features on the surface of Mars and that they formed a global network. Such a network of linear structures would be unlikely to form via natural processes, and he instead claimed that they were evidence of a Martian civilization.

Lowell expanded on Schiaparelli's work, producing detailed sketches of a Martian surface crisscrossed with a web of canals. It was troubling that other astronomers who attempted independent observations of Mars were unable to verify Lowell's claims. Lowell responded by questioning whether only the most powerful terrestrial telescopes located at the best viewing sites—and by that he meant his own—would able to resolve such features.

As a leap of faith, Lowell's claims remain formidable even to this day. What reason would a Martian civilization have for indulging in a spate of spectacular planetwide engineering? Consider that from Mars, even using present-day telescopes, one would be able to see few or no man-made structures on the surface of Earth (though if clever, you would detect the glow of our cities at night). Lowell speculated that only some great need would drive the Martians to build on such a spectacular scale. Assuming that the red color of the Martian surface derived from a dry, dusty, dying planet, he dreamed that the Martian canals were aqueducts delivering life-giving water from the frozen poles to the center of Martian civilization on the equator.

So far, so unscientific. Lowell had no evidence for these claims beyond the observation of fleeting linear features on the surface on Mars, glimpsed in what appeared to be brief moments of atmospheric calm. What changed? Telescopes changed. They increased in aperture and resolving power. The interplay of light and dark regions on the Martian surface and the blurring effects of Earth's atmosphere caused fleeting linear features. With larger telescopes and clearer views of the Martian surface, the canals faded like a dream upon waking. And thus Lowell's dream faded and died with him in 1916.

But Lowell's passion for astronomy and the observatory that bears his name left a more permanent legacy. In 1930, Clyde Tombaugh, working as an astronomer at the observatory, used photographic exposures of the outer solar system to follow a faint smudge of light in orbit about the sun. The smudge turned out to be Pluto, which at the time was, and for many still is, the ninth planet from the sun.[5]

Billions and Billions of Planets?

Do there exist worlds beyond the solar system? Looking skyward on any given night, your unaided eye will see roughly three thousand stars in the night sky. There will be another three thousand or so behind you, on the other side of Earth. All these stars are located in the Milky Way galaxy—our galaxy. With the aid of a telescope, more, apparently fainter stars can be seen. Although it is not possible to count each star within the Milky Way (they crowd together to the point where they are difficult to resolve individually), we can count the total light they emit and compare that figure to how much light we expect a "typical" star to give off. Such light counting reveals some four hundred billion stars within our galaxy. Each star is a sun very much like our own. Some are closer to us, some farther. Some are hotter and more luminous than our sun; others are cooler and fainter. Each is a luminous globe of ionized gas powered by a range of nuclear fusion reactions occurring in its core. In this sense, all stars belong to one family of objects.

Our sun is accompanied by a system of planets. Do the other stars in the Milky Way host similar planetary systems? Many astronomers from classical times onward have expected to find planets orbiting other stars. There appears to be nothing special about our solar system—created from the dust and gas left over during the formation of the sun. In addition there is nothing special about our sun—many stars of similar mass and composition appear to exist throughout the Milky Way.

Expectation does not equal discovery. It was only in 1995 that astronomers confirmed the existence of the first planet in orbit about a normal, or main sequence, star. The method they used was simple and elegant: although the light from the planet itself is lost in the much brighter glare of its parent star, the planet exerts a rhythmic gravitational tug on its stellar parent. Rather like two eternally mismatched dancers, the smaller planet waltzes around its stellar partner, which in turn is drawn on a much smaller yet complementary orbit. The scientific term for this measurement technique is the stellar radial velocity method, more commonly known as the Doppler wobble: as seen from Earth, a distant star appears to approach and recede from us in response to its unseen planet.

The planet discovered in 1995 is referred to as 51 Pegasi b. Its parent star is 51 Pegasi a, a sun-like star fifty light-years from Earth in the constellation Pegasus. The planet 51 Peg b orbits its parent star every 4.2 days and imparts a maximum Doppler velocity of 56 meters per second to its parent. For comparison, in our solar system, Jupiter imparts a Doppler velocity to the sun of 12 meters per second as it orbits every 12 years.

In a neat circle of logic, the same mathematics that Johannes Kepler—a contemporary of Galileo—used to describe the motion of planets in our solar system was applied to understand the 51 Peg planetary system. The results were surprising to say the least: 51 Peg b turned out to belong to a new class of planet that we now call a "hot Jupiter." The mass of 51 Peg b is likely half that of Jupiter (or slightly greater than half). But its short orbital period implies that it orbits its parent star at only one-twentieth the distance between Earth and the sun. Since 51 Peg a is a sun-like star, the surface of the planet (more likely, a high-altitude atmospheric layer) is seared to temperatures of over 1,200 Kelvin.[6]

What is important for our story is that in 1995 we first discovered an extrasolar world. Although observed by its indirect effect, we know that 51 Peg a is accompanied by a planet—one unlike any we ever imagined. The Doppler technique reveals the mass of the unseen planet and the distance at which it orbits its parent star. Furthermore, our observations of the parent star tell us how hot the surface of the planet must be. To put it in perspective, we know almost as much regarding any given extrasolar planet as we did about the outer planets in our solar system before the beginning of space exploration. Each is a world. We can measure its physical properties. We can assess the extent to which each is a potential habitat for life. We are even at the very beginning of being able to search for the signatures of life in such extrasolar worlds.

As of 2014, just over 1,800 planets have been discovered orbiting other stars—some occur singly, some in multiple systems. This number represents those planetary systems that astronomers consider to be "confirmed"—usually via measurement of a stellar Doppler velocity signature. Some few thousand others, in particular those discovered by the Kepler space mission, which we will meet in chapter 8, are considered "candidate" planets awaiting confirmation. Now the careful reader will note that this section is titled "Billions and Billions of Planets?" How did I get from 1,800 to billions? Not all stars in the Milky Way have been surveyed for planets, but of those that have, a remarkably high fraction appear to host planets. Astronomers refer to this number as f_{planet}, the fraction of stars of a certain type that host planetary systems. It turns out that for normal, or main sequence, stars, which make up the bulk of stars in the Milky Way, f_{planet} is somewhere between 0.1 and 1 (with 1 meaning that every such star hosts planets).

Hang on a minute—that is astounding. Astronomers are used to dealing with numbers that are, well, astronomical: numbers so large that we don't have names for them (the mass of the sun for example is 2×10^{30} kilograms—that's 2 followed by thirty zeros; the average density of matter and energy in the universe is 9×10^{-27} kilograms per cubic meter—which is 9 preceded by 26 zeros). In astronomer-speak, a number between 0.1 and 1 is basically 1. So within our order-of-magnitude approximation, a planet accompanies every star.[7]

If you were happy with my earlier statement that there is nothing remarkable about the makeup and contents of our solar system, then you should not be overwhelmed by this revelation. What is staggering is that as you look out on a starry night at those three thousand naked-eye stars in the Milky Way, you should expect each one to host a planetary companion. Many perhaps have their own planetary systems. None will be an exact copy of our solar system. Yet given the range of planetary masses and physical composition we expect to find, they are effectively identical. And when your imagination leaps to the four hundred billion stars that we believe make up the Milky Way,

ALIEN EXTRAVAGANZA

you should expect to find four hundred billion planets (or so) waiting there for you.

It's Life, Jim, but Not as We Know It

What will aliens look like? I think we all get the idea that aliens in movies normally look humanoid, for two reasons: it is cheaper to portray them that way, and humans like to anthropomorphize aliens. There are many notable exceptions to this generalization, but the serious question is, what should our starting point be for recognizing alien life?

While I am not ruling out that one day a Mars rover will capture a time-lapse image of a tiny Martian slug eking its way across a dusty plain, our search may well be subtler. The phenomenon we call life is a set of linked chemical processes, and the energy required by life generates a series of chemical by-products (breathe out and you will see what I mean). Therefore, our search for life would do well to consider how life processes alter the chemical makeup of a given environment. Chemical signatures resulting from biological activity are referred to as biosignatures. The best, or clearest, biosignatures are those that cannot be reproduced by nonbiological chemistry.

On Earth, the abundance of atmospheric oxygen produced by plant photosynthesis is a clear biosignature. Distant observers of Earth may well exercise some caution when they note that one-fifth of our atmosphere consists of oxygen—there could be some yet-undiscovered nonbiological process at work. But they would note that our planet differed from many others in a way that could point to the existence of life. We would certainly merit a closer look. This, then, is the sense in which astronomers define biosignatures (or in this case, an atmospheric biomarker).

So it may be more practical to first identify the signatures of life. But what of the agents of change, the organisms themselves? I stated earlier that when faced with questions of how to search for life in the universe, we should start with what we know of life on Earth and then see in what direction we can reasonably extrapolate this knowledge.

11

From this perspective, we should start thinking about the simplest organisms on Earth: single-celled bacteria and archaea.[8] Whichever way you look at them, these organisms dominate life on Earth. Bacteria and archaea, which form the largest source of living material (biomass) on Earth today, have existed continuously over the 3.5 billion–4 billion years during which there has been life on this planet (dinosaurs notched up some 165 million years; we are at 2 million years and counting).

The point here is to think simple—especially if you work in a laboratory constructing a suite of instruments to perform a remote search for life on a distant planet or moon. Your definition of success may well be to create an instrument that can detect a biosignature of alien life. Having done that (and collected your armful of scientific prizes), you can then think about how to work out what the space sludge actually looks like and what it is made of.

Contact

Where will we discover new life? Might we detect life signatures from a sample of bacterial goo on Saturn's moon Titan? Might observations of an exoplanet reveal an atmospheric biomarker? Could it be artificial life created in a test tube here on Earth? Or might we receive a personal message beamed in from a distant, intelligent life-form? All these are possibilities. The challenge as a scientist equipped with limited resources is to decide where your search for life should be directed. Put simply, if you can fund one space mission, where will you send it?

When I put the above questions to my students, the bulk opt for bacterial goo or biomarkers. A few go for test tube life, and one or two are patiently waiting by the phone. Their answers mainly reflect their backgrounds as scientifically literate university students. The aim of the question is to get them to think about scientifically concrete contact scenarios and how we will react—personally and scientifically—to them.

Now for an interesting one—when will we discover new life? In ten years? A hundred? What about a thousand? Once again the answer depends on your outlook. Ten years is probably optimistic.[9] It essentially assumes that life is abundant in the locations where we are already investigating and that we have deployed equipment capable of recognizing it unambiguously. One thousand years could be called pessimistic. It essentially places discovery in some distant, future generation so far removed from our present-day efforts that one assumes the near-term chance of success is zero.

The answer of one hundred years is more interesting. Its span of several tens of years is of the order of a human lifetime. It is also of the order of the time required to conceive, construct, fly, and learn from a robotic space mission to Jupiter or Saturn. It is of the order of the length of time to construct the next generation of massive telescopes required to probe the atmospheres of distant exoplanets (thirty-meter-diameter mirrors are currently under construction). One hundred years appears to be achievable—if. If we make informed decisions. If we remain bold in our attempts. And if we get lucky (we will come across a few examples of bad luck in our story).

A Journey of a Trillion Miles

With the discovery of an abundance of new worlds orbiting distant stars, and with the increasing scientific scrutiny to which a host of space probes are subjecting the worlds of our solar system, the search for life beyond Earth is in the throes of a revolution—one every bit as exciting as that brought about by the application of the telescope to astronomy. Our knowledge is expanding at a staggering rate, and yet with the absence of any secure detection of life, it remains incomplete.

The aim of this book is to push modern astrobiology to focus on five plausible scenarios for the discovery of alien life. Why five, I hear you ask? Mainly because "plausible" is not the same as "likely." If I tried to tell you that there was one direction, one planet or moon, where we would be most likely to discover life beyond Earth, then I would be presenting you with spin rather than science. Alternatively, if I presented you with an exhaustive list of alien-contact ideas, I would be ignoring the practical reality that we can fund and pursue only a limited number of scientific endeavors. So by focusing on five scenarios for the discovery of alien life, we hope to strike a balance between the above two extremes.

The fact that we are even able to speculate on the types of living organisms that we might find in specific locations in the universe is a testament to how the science of astrobiology has progressed and matured over the last twenty years. Is such speculation justified, or is it instead a flight of scientific fancy? To answer these questions, it is worth considering that any new scientific experiment requires a certain amount of speculation: if you knew exactly what to expect, why would you perform the experiment?[10] The kind of speculation undertaken in this book is essentially identical to that performed by a team of scientists planning a major new space mission.

The team that sent NASA's *Curiosity* rover to land on Mars in August 2012 did not know exactly what it would find. Previous Mars missions had discovered a great deal of circumstantial evidence for surface geology influenced by the action of liquid water. *Curiosity* was dispatched to Mars with a suite of instruments that promised new discoveries based on our total knowledge of Mars accumulated up until that date. In the future, new, yet-unplanned Mars missions will seek out the traces of specific organisms in particular Martian habitats. If the mission is robotic, the technical team will have had to speculate on the types of life they would expect to find, and to design experiments likely to identify it. If their speculation is well founded—and life exists—their mission stands a reasonable chance of success. If their reasoning is flawed, or if life does not exist (or they are unlucky), then disappointment looms.

Our premise is therefore that for at least five scenarios featuring the search for alien life, we have accumulated enough scientific data to make what we hope is informed speculation about what kind of life we might discover. A great deal of uncertainty undoubtedly remains—and indeed is part of the thrill of the challenge. We are free to make decisions—good or bad, informed or uninformed—that may lead us either to success or to frustration. There can be no better parting words than those of Giuseppe Cocconi and Philip Morrison, who in 1959 challenged the scientific community to search for extra-terrestrial intelligence: *the probability of success is difficult to estimate; but if we never search, the chance of success is zero.*

The properties of the universe provide quite literally the biggest vantage point from which to frame our questions about life within it. How do the age and size of the universe affect the search for life? Where did the ingredients for life on Earth come from—and are they found elsewhere in the universe? How early in the history of the universe could life have arisen? If our solar system had formed much earlier, could Earth have formed? Could life have appeared? And where exactly in the universe are we capable of searching for life: our local stellar neighborhood, the Milky Way, the whole universe?

The Night Is Dark, yet Full of Stars

How big is the universe? Has it existed forever or for some shorter, limited time? The answers lie in the sky above. Have you ever looked at the night sky and wondered not so much what is the nature of the stars and galaxies you see, but why there is dark sky between them? The question of why the night sky is dark is most often called Olbers' paradox, and by taking the paradox apart and then reassembling it using a manual inspired by modern cosmology, we can learn some profound truths regarding the universe. Put simply, the universe may have no physical limit in space. It could be infinite. It began at a particular time, though, and therefore has a finite age. Together with the fact that light travels at a fixed speed, the finite age of the universe means that at any time we can see only the limited portion of the universe from which light has traveled to us. We call this limit the cosmological horizon, and it defines the observable universe as something distinct from the unseen portion of the universe beyond the horizon.

Right now you would be justified in requesting a little more explanation and little less profound leaping. So how does the darkness of the night sky lead us to the idea of an observable universe of well-defined age? Heinrich Olbers (1758–1840) and his peers imagined a universe of unlimited extent with an even scattering of stars in all parts. Whichever direction you looked—anywhere in the sky—your gaze would fall upon a star. Some would be closer, some more distant, yet if the universe were also infinite in age, we would be able to see them all.

Wouldn't the more distant ones be fainter? Quite right. Good point—which gets me onto how astronomers describe the tendency of objects to get fainter the farther they are from us. One way of achieving this is to describe the stars by their surface brightness: their brightness divided by their apparent size. Imagine they have the same surface brightness as the sun. The sun appears to us to be about half a degree across in size.[1] If you multiply the surface brightness by the apparent size, you get the real, total brightness of the sun. If we move the sun farther away, the surface brightness stays the same; it is the apparent area that decreases. Objects appear smaller the more distant they are. A very distant star has the same surface brightness as a nearby star; it just appears to be much smaller and therefore appears fainter. But what if every direction we look in ends on a star, each star having the same surface brightness? In this case, the small apparent sizes of the distant stars would not matter. If the small circles that represent the disk of each star overlap, then the surface brightness of the sky in that direction would be the same as the surface brightness of an individual star. The sky should look like the surface of one big star.

But it doesn't, though, does it? We must have taken a wrong turn somewhere. So if we retrace our steps, what could we do differently? We could make the universe finite in space by adding an edge. We could put stars in only some parts of the universe but not everywhere (we could also require that more distant stars have lower surface brightness, etc., but I think you get the point). We could make the universe finite in time so that light has had only a limited amount of time to reach us—we would not be able to see the whole universe. Astronomy in the nineteenth-century world of Olbers could not provide a definitive answer. A new view of the universe was required.

The Realm of the Nebulae

In 1929, Edwin Hubble published a series of observations that appeared to show that the universe was expanding away from us, in all directions, all at the same speed. Essentially, we appeared to be at the center of a great explosion of galaxies. He noted that the brightness of reference stars in his sample of galaxies was linked, or correlated, with an apparent velocity shift measured from a dispersed, spectral image of each galaxy. Since brightness is related to distance, and the recession velocity produced a spectral shift to the red, we refer to Hubble's discovery as the distance-redshift relation. So what could be so special about our location that the rest of the universe is rushing away from us? The story of modern cosmology is intriguing because it features threads, contributed by both theorists and observers, that are at best only loosely intertwined. Thus, in work largely overlooked by Hubble, two European cosmologists had effectively translated Einstein's general theory of relativity into a language that could describe the universe itself.[2]

Alexander Friedmann, who put his knowledge of ballistic mathematics to use as an artillery observer during the First World War, had shown in 1922 that a universe described by Einstein's theory would be a dynamic entity: its physical size would either increase or decrease as time passed. In fact, the more difficult thing to achieve mathematically was to make it stand still. In 1927, Georges Lemaître explained how Friedmann's work on an expanding universe would lead naturally to exactly the kind of relationship between distance and redshift that Hubble was to confirm two year later.

The subtle point in Lemaître's explanation was that expansion occurred in the entire fabric of the universe—what we call space-time—and that galaxies would be carried along for the ride. An observer in one galaxy would see all others recede from him. And an observer in the next galaxy over would see exactly the same effect. Universal expansion has the giddy consequence of making each galaxy appear to be at the center of all the action. Furthermore, Lemaître understood that if one were to mathematically run the clock backward, the galaxies that populated the universe would approach each other until, at some point in the past, they would meet. Within his mathematical model, this point represents the beginning of the universe as we understand it—what would later be called the big bang. The time between the big bang and today is the age of the universe.

Hubble wrote down the distance-redshift relation in the following way: galaxy recession velocity=H×(distance to the galaxy). The value of H in the universe today is what astronomers refer to as Hubble's constant. We can use Hubble's formula to ask an interesting question: if the recession velocity of a galaxy has not changed much over the age of the universe, by how much would we have to run the clock back until all galaxies merged into one primeval lump? Well, time is just distance divided by velocity, so if you use Hubble's formula linking the two, you find that the time taken is $1/H$. The modern value of Hubble's constant is pretty close to 70 kilometers per second per megaparsec (the units are not that inaccessible; what it means is that if a galaxy is one megaparsec away from us—3.26 million light-years—it will appear to recede from us at 70 kilometers per second). If, in addition, you take account of the fact that the expansion velocity of galaxies has actually changed a little over the history of the universe, this value of Hubble's constant tells you that the age of the universe today is 13.8 billion years.

So not six thousand years, then? No, not really. It is worth noting that when the work of Friedmann, Lemaître, and Hubble was combined in the 1930s, it marked the first scientific determination of the age of the universe. At the time, inaccurate measurements resulted in a value more like one billion years instead of the value we accept today. When the values from astronomy were combined with the first radiometric dating of Earth rocks and calculations of the lifetimes of stars derived from the new science of nuclear physics, a startling and coherent picture emerged: Earth, the stars, and the universe as whole were older than we could ever have imagined. Not thousands or millions of years, but a thousand times older.

Putting aside this scientific jolt of the first magnitude to our collective consciousness, what remains is our solution to Olbers' paradox: the universe began at a fixed point in time. The most distant light we can observe today has taken 13.8 billion years to reach us. This is the observable universe. What lies beyond? Possibly a great deal. The universe may be infinite—we just don't know. The light is en route to us, and as time passes, the observable universe keeps getting bigger, but only one year at a time. I hope you like the view—because it is not going to change any time soon.[3]

The Clock Is Ticking

So what has the universe achieved in the last 13.8 billion years? Where does life fit into this great cosmic time frame? One of the most elegant and accessible ways to approach the history of the universe is to reduce the span of cosmic time to one year and follow events on a cosmic calendar. The big bang occurred in the first moment of January 1. You are reading this book in the dying moments of December 31. What highlights can you look back on over the past year?

In its first moments, the universe expanded as pure energy and then particle physics took over and a host of fundamental particles broke loose like so many liberated zoo animals intent on a cosmic rampage.[4] The material of which you and I are made started out as a relativistic soup of fundamental particles. As the universe expanded, the soup cooled. Atomic matter—more familiar to us today—formed from the fading glow of the big bang: hydrogen, helium, a pinch of lithium, and their isotopes. This atomic material then coalesced into the first cosmic structures: slowly cooling clouds of gas.

The first stars formed at the end of the first week of January, some few hundred million years after the big bang. The first galaxies appeared soon after, yet one has to wait until March—ten billion years ago—before the Milky Way galaxy was largely assembled. August—five billion years ago—was a good month for us: the sun formed, quickly followed by the planets of the solar system. In September, just a few hundred million years after the formation of Earth itself, we have the first tenuous evidence of life on Earth—simple, single-celled life-forms.

Only in November, some two billion years later, does life overcome its initial simplicity and follow a path to more complex, multicellular life. When December comes around, the highest form of life on Earth is basically slime. Then something interesting happens. For reasons that we can only speculate about, the conditions on Earth and perhaps life itself reached a critical point: an explosion of evolutionary activity created a multitude of complex life-forms. This event happened on December 15, roughly 540 million years ago, and is known as the Cambrian explosion from its location in the geological record.

Life was very much on the first few rungs of the evolutionary ladder. Dinosaurs appeared on Christmas Eve, and our first mammal ancestors on Christmas Day. The dinosaurs overdid it rather at Christmas and were wiped out by an asteroid impact on December 29—sixty-five million years ago. If nothing else, this event may have cleared the evolutionary playing field to the extent that some fairly insignificant mammals were able to evolve into newly vacated niches. Much like modern humans, only on New Year's Eve did a group of mammals finally get their act together and start thinking about the future. At 10:15 a.m. the first apes appeared, learning to walk upright only at 9:24 p.m. after a staggering eleven hours (or seventeen million years) of evolutionary effort. We learned to write fifteen seconds ago, yet managed to build the pyramids only five seconds later, which is not bad going. Finally, rather like some marathon runner in a gasping dive for the line, Christopher Columbus made it to America just one second before midnight on the eve of a New Year. Take a deep breath—you've just run a long way.

If we played the whole sequence of cosmic history again, could life have arisen earlier? Should life on Earth have done better? Or are we already high achievers, bucking some universal trend against complex life? What might have happened if we had changed the conditions in the universe a little bit?[5] Would life have evolved differently? Should we even be focused on complex life—isn't simple life, even if different from that on Earth, interesting enough? It depends on what you want to do with it: discuss movies or learn about life processes. To understand under what conditions life could or could not arise in the universe, we need to consider what living organisms are made of and where that material comes from.

We Are All Star Stuff

The periodic table of the elements is a work of art. It is also the most successful scientific diagram ever created.[6] It describes every known chemical element and reveals in stunning clarity the patterns arising from the hidden structure within each atom. The elements of the periodic table form a sequence: each element is described by its atomic number, which in reality is just the number of protons it contains. Hydrogen has one proton, helium has two (neatly balanced by two neutrons as well), lithium has three, and so on. It permits us to answer some basic questions about nature (so basic you may have never thought of them): Is there an element lighter than hydrogen? Does an unknown element exist between hydrogen and helium? The answer to both questions is no—you can't make an atomic nucleus using a fraction of a proton. There are no gaps in the periodic table: we know of every element in nature up to uranium (ninety-two protons). We even know the sequence of elements heavier than uranium—the wonderfully named transuranic elements. These are short-lived, radioactively unstable elements created and studied in nuclear laboratories.

Where did all these elements come from? Is Earth favored in some way, having collected a full set? Let's go back to the beginning: if you started counting at the moment of the big bang, by the time you reached two hundred or so, the observable universe would be about a light-year across. All of what we consider "normal" matter—protons, neutrons, electrons—would be in the form of a plasma at a temperature of several million degrees. The age of primordial nucleosynthesis, a brief and early phase of nuclear fusion spread throughout the young cosmos, would have just ended. Depending on your perspective, it wasn't terribly productive: roughly 25 percent of the hydrogen permeating the universe fused to form helium. A smattering[7] of the helium fused a little more and formed lithium. That was it. From the first few minutes of universal existence until about six hundred million years later, no new atomic elements were created.

While this may seem like a shamefully long break to take after a period of energetic though limited activity, there was a good reason for it. Nuclear fusion occurs only under conditions of tremendous temperature and density.[8] These conditions occurred for a few minutes in the early history of the universe. They would not occur again until the first stars formed. Within the cores of the first generation of stars, the conditions of temperature and density once again reignited the fire of nuclear fusion.

Stars are essentially nuclear pressure cookers; elements fuse together in a nuclear riot, producing heavier and heavier atomic nuclei right up to iron (with twenty-six protons). The particulars of nuclear physics mean that the fusion of nuclei lighter than iron generally produces a little burst of energy over and above the hot temperatures required to slam the two nuclei together. This energy keeps the plasma hot and allows more fusion to occur. But beyond iron, each fusion event consumes a little bit of energy and cools the nuclear house party. The result is that stars, especially massive ones, are effective at producing elements up to iron, but not beyond.

That is barely one-third of the way along the period table. How will we get to the end? At the end of a star's lifetime, when the great pressure exerted by its outer structure is no longer able to force the fusion of nuclei in the core, a cataclysm can occur. Stars of low mass—up to a few times the mass of our sun—end their lives as a white dwarf, a stellar ember that was once the hot core of the star. The fire of fusion died, and the hot coals slowly (actually, very slowly) cool and fade.

A very different fate awaits stars of higher mass. A white dwarf is supported against gravity by a quantum exclusion principle that prevents the electrons in the dead star from packing too closely together. We call this effect degenerate electron pressure. In more massive stars, this force cannot withstand the tremendous gravitational pressures. The dead core of the star collapses to become a neutron star only a few kilometers across and supported by the degenerate pressure of neutrons rather than electrons.[9] To put it in perspective, the radius of the sun is some 700,000 kilometers, or about 100,000 times as large as the radius of a neutron star. As the outer layers of the dying star collapse under gravity onto the neutron star, a tremendous nuclear flash occurs as the temperature and density of matter spark a last, all-consuming burst of fusion. Elements all the way up to uranium (and possibly beyond) are created and then violently ejected into space. You have just witnessed a supernova.

Though it may not seem like it, supernovae play a fundamental role in the story of life in the universe: they complete the nuclear production of the periodic table and act as the delivery mechanism, seeding their local neighborhood with a rich variety of new elements. All the elements beyond hydrogen and helium that surround you now, and all those present in life processes—from the iron in hemoglobin flowing through your blood to the magnesium atom at the center of a molecule of chlorophyll—can trace their history through the core of a star and the blast of a supernova.

Our Place in the Cosmos

So the universe is 13.8 billion years old. It is very large, quite possibly infinite in size, and the multitude of atomic elements have been produced in stars and supernovae and then dispersed throughout space. What I now want to consider is how Earth, our sun, and its solar system fit into this vast cosmic landscape. How did they come to be, and was it the result of a rare piece of luck or a common event?

We left the story of our expanding universe shortly after the formation of the first stars and supernovae, a few hundred million years after the big bang. As if in a reflected response to the expansion of the universe, matter began to contract under gravity. Gas clouds collapsed and collided, swirling to form the early ancestors of galaxies. Lit from within by new generations of stars, the galaxies became the immense cities of stars that we observe today. The space between galaxies is vast, and only rarely do massive galaxies collide with each other in great cosmic train wrecks. For most of their history, they are isolated and aloof. Within each stellar city block, generations of stars are born out of clouds of gas and dust, each star shining with the energy liberated by nuclear fusion, each nuclear reaction populating a new space on the star's individual periodic table. The fate of massive stars is to explode, dispersing their heavy elements, which slowly accumulate in subsequent generations of stars.

Our sun was born into one such later generation. Even after several generations of stellar nucleosynthesis, the gas cloud that collapsed to form our sun had been enriched with heavy elements[10] only to roughly 2 percent of the mass of the cloud. You might reasonably ask why this number is so low—why not 10 percent, 50 percent, or even larger? The answer is that most of the material of which a star is made never undergoes nuclear fusion. The role of the outer layers of a star, those outside the core, is to provide weight. The weight of material pressing onto the core creates the conditions of temperature and density required for fusion to occur. Once the fuel composed of lighter elements in the core has been depleted, the star will evolve either to become a white dwarf or to go supernova. The rest of the star is ejected into space.

Our sun and solar system began, therefore, as a slowly tumbling cloud of enriched gas. As the cloud cooled, it collapsed on itself, spinning faster all the time and taking on the appearance of a flattened disk. Gas in the center piled up on itself until the central temperature and density sparked fusion into life and our sun was born. The flash of light thus liberated lit up the gaseous outer parts of this disk and blew away much of the material. The dust of heavy elements that remained started to coalesce, slowly at first, into microscopic grains. Small electrostatic charges on the grains attracted more grains into clumps. The clumps grew under gravity into rocky lumps, colliding, destroying, and sometimes adhering to one another in some vast chaotic round of bumper cars.

The winners of the round grew into the planets we know today, each growing planet devouring material in its orbital path like some voracious cosmic predator. The outer planets chanced to grow large, capturing large quantities of cold gas from the frigid outer solar system to become the gas giants we know today—Jupiter, Saturn, Uranus, and Neptune. The inner solar system was baked free of volatile elements by the heat of the young sun, and remained the preserve of the rocky planets—Mercury, Venus, Earth, and Mars. The rest is just debris: the asteroid belt between Mars and Jupiter, the Kuiper belt beyond Neptune.[11]

Is this story unique to our solar system? In specifics, yes: the particular arrangement of planets is probably unique to our solar system. In general properties, no: we observe swirling clouds of gas and dust around many young stars. Some present the appearance of a disk; some even show clumpy orbital lanes that appear to be sites of planetary growth. Our view is therefore that planets form with their young stars from gas enriched with heavy elements. Astronomers refer to such heavy elements generically as "metals." Rocky planets and the cores of giant gas planets are made from these metals. Rather like the raw materials for building a house, more metals may result in more planets or a wider range of planets. At this point, we just don't know. We know that since metals are required to build planets, more is better than less, but beyond that our knowledge of the specifics of planet formation is patchy at best.

Could Earth have arisen earlier in the history of the universe? With each passing generation of stars, the metal enrichment of interstellar gas increases. We don't know whether there is a specific threshold of metals required for planet formation. In general, there were fewer metals earlier in the history of the universe, so it may well have been more challenging to scrape together the materials to make planets. But localized clumps of massive stars will quickly explode and rain a debris of heavy elements down on their nearby patch of space. So at this point we can say only that it would probably have been harder to make planetary systems earlier in the history of the universe, but with the sometimes infuriating tendency of scientists to hedge, we cannot rule it out. What we can say, though, is that wherever there are stars, there will be chemical enrichment. Though not quite yet the stuff of life, the raw materials for complex chemistry are abundant throughout the Milky Way and throughout each galaxy in the present-day universe.

Beyond the Milky Way

The distance to the nearest star system—Alpha Centauri—is 4.3 light-years. The distance to the center of the Milky Way is 26,000 light-years. The nearest galaxy to us is Andromeda, and it is 2.5 million light-years from Earth. If you go out tonight and look toward the faint smudge on the sky that is Andromeda, the starlight you observe is 2.5 million years old. It left Andromeda when your early human ancestors were learning to use stone tools. Together, Andromeda and the Milky Way form a small clump of galaxies known as the Local Group. Beyond this, however, the typical distance between large galaxies is 15 million light-years.

The distances between galaxies are incomprehensibly vast. Even when one day we fly our first space probe to the nearest stars, the idea of traveling to the nearest galaxies will remain in the realm of science fiction. In addition, many of the observational efforts described in later chapters, hunting for planets via their stellar Doppler signatures or via repeated planetary transits, for example, are ineffective over such tremendous distances. We will learn about the search for extraterrestrial intelligence (SETI) and the longing for interstellar conversations. But when compared to the travel times of light even within our own galaxy, 15 million years seems like a long time to wait for someone to pick up the phone.

The reality of the present-day search for life in the universe is that, if the solar system is our local street and the Milky Way the city in which we live, our search for life will be limited to sending space probes along the road outside our house and performing remote observations of the blocks of the nearby stellar suburbs. The other galaxies really are far, far away.

To Infinity and Beyond

I hope you can now appreciate the remark I made in the first chapter: in an infinite universe, all things, including life, are not just possibilities, they are certainties. Each galaxy contains hundreds of billions of stars. Each one may be accompanied by a planetary system. Beyond the Milky Way, the observable universe may contain hundreds of billions of galaxies. That alone gives a possible 10^{22} planets. But it is the possibility of an infinite universe beyond the cosmological horizon that takes the number of possible habitats for life from the merely numerous to the strictly infinite. But let's remember that such numerical gymnastics brings us no closer to examples of life beyond Earth. We must start somewhere. So let's start with the solar system.

three WHAT ON EARTH IS LIFE?

What on earth is life? What is life on Earth? At present, these two questions are inextricably linked. The only life we know is that of Earth. It defines all our knowledge of life—at least all that can be tested by experiment and observation. Only by discovering new life, either created by ourselves in an Earth laboratory or by searching beyond Earth, will we develop a deeper understanding of the nature of life. One principle, though, appears to be clear: the emergence and evolution of life on Earth is closely linked with the corresponding formation and evolution of Earth itself. Since early times, the changing physical conditions on Earth have influenced the nature of life, and vice versa.

We therefore have our work cut out in this chapter: Can we agree on a definition of life? What exactly is life, and at what point does chemistry become biology? How did conditions on the early Earth lead to the emergence of life? What do the fossil record and geochemical evidence from ancient rocks tell us of the development of life on Earth? Finally, how has evolving life changed the conditions on Earth, and how has the planet sustained and regulated the life it hosts?

But more broadly, as astrobiologists, we have to ask how much of this is applicable to our search for life beyond Earth. What general principles will prove valuable? What specific insights will lead to a narrowly defined search? Was the emergence of life a natural consequence of the physical conditions of the early Earth? If so, is it reasonable to expect life to arise on all planets that possess the same broad properties of the early Earth—for example, a temperate atmosphere surrounding a solid surface covered with liquid regions rich in organic chemicals? Once established, what properties of a planet determine whether life will persist? Which planets offer only fleeting habitats for life, which provide long-term stability?

A Pause for Reflection

If I asked you to look in a mirror and describe yourself, how would you answer? Would you reply that you are a living creature and that you share some basic characteristics with all living creatures found on Earth? Think of yourself this way: your body's biochemistry is organized into cells, you are the product of reproduction and evolution, and you have grown throughout your life. You daily exhibit a flexible metabolism; you convert fuel into energy that powers your body while regulating your bodily functions to stick within well-defined operating limits. Although my portrait may have failed to capture your personal idiosyncrasies, if you were to attempt to define life in its most general sense, you would probably have to resort to something similar to the above description.

This rather verbose answer to what appears to be the simple question "what is life?" is a reflection of the fact that life is a phenomenon rather than a simple quantity. I could describe myself as being six feet three inches tall and weighing [redacted in print version]. But I could not describe myself as having 1.73 units of life. An interesting approach might be to describe myself as being composed of approximately ten trillion cells, although different living creatures possess cells of differing levels of complexity and activity. I think you see my point: life is a series of linked phenomena that defy simple measurement.

All the above is very good for affirming your sense of being a worthy addition to the cosmos, yet does it bring us any closer to one of the main goals of astrobiology—recognizing the presence of life? You could argue that yes, the above ideas are useful. If we could identify a sample that is organized into ordered structures, displays chemical evidence for a positive energy cycle, and possesses a mechanism that both encodes the overall chemical recipe of the structures and permits them to reproduce, then many researchers would agree that the discovery shares sufficient characteristics with life on Earth to be considered life in its own right.[1]

This is a theme we will return to later in the book when we travel beyond Earth to the planets of our solar system and consider the kinds of scientific tests required to provide unambiguous answers in our quest for life. At this stage, I hope you will agree that we can all benefit from a little personal reflection in our search for a definition of life.

This Island Earth

Planet Earth has been likened to the best spaceship we will ever possess: it carries an abundant cargo of raw materials, its living systems convert solar energy into usable fuels, and it shields us from the harshness of space. Life has existed on Earth for close to four billion years, and the stability of the surface environment over that period has played a critical role in permitting life to evolve into its modern forms. It is interesting to consider how the physical properties of the planet—its geology—have provided us with a stable habitat and protected us from harm.

Earth's atmosphere was originally created, and is replenished even today, by volcanic outgassing. The chemical composition of the atmosphere provides us with a weak greenhouse effect—a thermal blanket that traps solar radiation—and warms the surface.[2] Besides warming the planet, our atmosphere provides enough pressure for water to exist as a liquid over most of its surface. To get a sense of what the conditions on Earth would be like in the absence of an atmosphere, just look at the moon. Take away the warmth, and the water would freeze; lower the pressure, and it would boil. Either way, our atmosphere is the key to surface life on Earth.

The atmosphere is protected from space—in particular from the energetic particles that make up the solar wind—by Earth's magnetic field. Unable to penetrate the magnetic field, the solar wind swirls around Earth, brushing only the topmost levels of our atmosphere at the magnetically exposed poles, where charged solar particles bring the aurorae to life. Without the protection of Earth's magnetic field, our atmosphere would be steadily ionized, ablated, and evaporated by the energy of the solar wind and would blow away into space.

By listening to the echoes of powerful earthquakes as they resound within the interior of the planet, we can discern the source of Earth's magnetic field. Seismology has revealed Earth's inner structure: a liquid outer layer composed of nickel and iron surrounding a solid inner core. The hot outer core is like a great beating heart where liquid metal circulates in giant convection currents. These currents drive great flows of electrons within the core of our planet—quite literally, an electric current. On top of all that, our planet rotates, whirling the electrical eddies at its heart into a vast magnetic field. It is this combination of a liquid metallic core and Earth's rotation that creates a planetary dynamo and drives our magnetic field. Both are critical: if you reduce the rate of rotation (as on Venus) or cool and solidify the core (as on Mars), the resulting magnetic field is dramatically weaker.

All this is driven by geology—the dynamic processes that result from Earth's hot, molten interior. It is amazing to realize that the solid rock that forms the foundation of our lives on Earth is nothing more that a few thin flakes of crust riding atop a bubbling globe of magma and metal. The heat within Earth is the still-molten glow of the energy released from the multitude of grinding collisions that created our planet 4.5 billion years ago.[3] Like some great heat engine, the hot core of our planet drives the geological processes that continue to power spaceship Earth to this day.

In the spirit of maintaining our open-minded perspective as budding astrobiologists, I am not trying to say that geological activity on a planet is either favorable or unfavorable for the possibility of life existing beyond Earth. The above relationships between geology and life do not mean that we should target exclusively geologically active worlds in our search for life. But they help us understand—in very broad terms—how geology on Earth has created and sustained the conditions in which life has evolved and thrived. If we imagine discovering extrasolar planets and moons displaying geological processes similar to those on Earth, we can now speculate on some

WHAT ON EARTH IS LIFE?

of the ways in which they might lead to or sustain life. But, as ever, we retain the right to be surprised by new worlds where oases of life may exist in geological deserts devoid of such activity.

Darwin's Triumph

Present-day life on Earth is both diverse and complex in its outward forms. We see examples all around us of incredible specialization and adaptation: if you were watching a nature documentary right now, at this point there would be a slow-motion montage of images showing hummingbirds, lions, antelopes, tree frogs, and other exciting creatures.

If we change our perspective and view life as a series of linked biochemical processes, we see that all—and I really do mean all—life on Earth shares an astonishingly small set of fundamental characteristics: the basic unit of life is the cell—a small packet of slightly salty water stuffed with enough organic chemistry to fill several textbooks. What is striking is that every cell appears to be following a fundamental set of common recipes taken from the same first-year biochemistry textbook.

All life encodes its genetic information by using deoxyribonucleic acid. DNA is essentially a long chain of basic molecular letters (referred to C, A, G, and T) linked together into an elegant double helix. It is like a molecular data bank containing all the information required to operate an organism's life chemistry, in addition to providing the mechanism to pass that information on to the next generation. On a similarly specialized level, the same unit of energy currency powers chemical transactions within each cell: exchanges use a particular family of phosphate groups based on adenosine triphosphate, or ATP.

The language of life's protein biochemistry is written with an alphabet consisting of twenty amino acids. The intriguing point here is that some five hundred amino acids are known to occur in nature, so why does life share this fundamental yet limited subset? A subtler shared characteristic of life is expressed in the uniform

33

chirality—essentially, the molecular shape or "handedness"—of these amino acids (and sugars). Given the tremendous diversity of modern life-forms, how did it come to be that this simple set of recipes left its mark on all life on Earth?

These two contrasting characteristics of life on Earth—the diversity of outward forms accompanied by a shared biochemical recipe—are reconciled by cell division and evolution. Cells reproduce by dividing themselves into two near-perfect copies. Natural selection seizes on each small variation in genetic properties and winnows out those better adapted to their environment like so much wheat from chaff. The complexity of outward form merely shows the action of evolution in causing us to adapt to the complex and varied environments we encounter on Earth.[4]

But what happens if we run the movie backward? We see complex life being slowly stripped away of its modern innovations. An ancient single-celled organism comes into view. Its biochemical architecture is simple yet robust: a DNA-based genetic alphabet, a metabolism powered by ATP, and protein chemistry expressed by a set of twenty amino acids. Through the shared biochemical recipes of modern life, we gain a fleeting glimpse of an early common ancestor to all life on Earth. It is an astonishing and profound realization: we are linked—by a single and continuous thread of cellular division and evolution—to the earliest moments of life on Earth. The persistence of this basic biochemical recipe for life indicates that from the earliest times, life got the fundamentals right and stuck to them.

Darwin himself realized this vision when he speculated in a letter of 1871 to his friend Joseph Hooker that life may have begun "in some warm little pond with all sorts of ammonia and phosphoric salts,— light, heat, electricity etc. present, that a protein compound was chemically formed, ready to undergo still more complex changes." This great leap of insight, which in one vaulting moment perceived life's origins across the deepest gulfs of time, was only one among Darwin's many triumphs. It has only taken us some 140 years of sustained scientific effort to keep pace with Darwin. Yet we can now pick up the thread of life's development and follow it, by using the fossil record and the atomic chemistry of Earth's rocks, back to the very earliest epochs of Earth's history.

The Realm of Hades

Geology tells the physical story of planet Earth and in many ways is the older sibling of evolution. Charles Lyell published his defining book, *Principles of Geology*, in 1830. His central thesis was that Earth's geological history was written over unimaginably vast intervals of time by the slow, steady action of forces that could be seen—if one had the eye to note them—acting on Earth today. The young Charles Darwin took volume one with him in 1831 when he sailed on the voyage of the HMS *Beagle* (he had to await volume two by post), and Lyell's vision made a deep impression on his own scientific view of the world. Evolution and geology have therefore grown up together: geology has revealed Earth's long history and provided the deep interval of time required for small evolutionary changes in generations of living organisms to produce the diversity of present-day life.

Modern geology divides Earth's history into four major eons: Hadean (the oldest), Archean, Proterozoic, and Phanerozoic (the youngest). The close link between the history of Earth and the history of life on Earth can be gleaned from the fact that the eons of Earth's history are divided according to the properties (or absence) of life as determined from the rocks of each period.

The Hadean eon covers the earliest epochs of Earth's history. The oldest geological data for planet Earth come from tiny crystals of the mineral zircon—the oldest of which possess radiometric dates as old as 4.4 billion years.[5] These tiny crystals, which represent mere fragments of the surface of the early Earth, are often found as ground-up debris located within still-ancient yet ultimately younger rocks. The Hadean Earth was young: its thin crust had only just begun to solidify over the molten interior. The surface of the planet was ravaged from within by rampant vulcanism and pulverized from above by a constant rain of protoplanetary debris—what we today call asteroids and comets.

35

The traces of this ancient bombardment have been erased from Earth's surface by 4 billion years of geological activity (vulcanism, plate tectonics, and good old-fashioned erosion). But the scarred history of this distant epoch is preserved in all its glory on the surface of the moon. The changing density of craters on the lunar surface, from the oldest highlands (dated radiometrically from rocks returned by Apollo 16 and 17) to the only slightly less ancient maria (Apollo 11, 12, 14, and 15), tell us that the bombardment lasted until some 3.9 billion years ago and culminated in a ferocious late burst of impacts. This, therefore, is the Hadean epoch of Earth—brutal, hellish, and exceedingly harmful to life.

The Archean: Echoes of Ancient Life

The Archean eon began 3.9 billion years ago in the relative calm that followed the end of the heavy bombardment. Indeed, the oldest crustal rocks on Earth date from only slightly before this period and provide firm evidence for a solid terrestrial surface. Some of the earliest evidence of life on Earth is hidden deep within the rocks of the Archean: fragile microfossils dating from 3.5 billion years ago that show fleeting details of individual cells. Individual microfossils are invariably found in layered macrofossils—the wonderfully named cryptozoon ("hidden life"): these mineralized remains of ancient microbial colonies are strikingly similar in form to present-day stromatolites, eerily primitive associations of bacteria and archaea perched on stool-sized outcrops in shallow saltwater lagoons.

Such microfossils represent the only visual evidence for Archean life. They show us an ancient world populated by colonies of exceptionally simple single-celled organisms. The cells themselves are primitive, or prokaryotic, structures that predate the evolution of the cell nucleus, and consequently their DNA, the genetic code on which they are based, floats freely in the main structure of the cell.

A neat and pervasive feature of Earth life permits a subtler, geochemical test for the presence of early life on Earth. All present-day life on Earth relies on the diffusion of carbon through a cell membrane, and the carbon then reacts in the aqueous interior of the cell. Two atomically stable forms of carbon exist in nature: carbon-12 (^{12}C—containing six protons and six neutrons) accounts for approximately 99 percent of naturally occurring carbon; the remainder is carbon-13 (^{13}C—six protons and seven neutrons).[6] The slightly heavier isotope ^{13}C does not diffuse through the cell membrane as readily as ^{12}C, and thus a lower fraction is taken up into the structure of the organism. Living creatures thus act as effective filters of carbon isotopes—and when ancient living creatures become ancient dead creatures, they become marine sedimentary rocks. Therefore, if you compare the carbon-isotope fraction of marine sediments with that found in other marine rocks (marine carbonates, to be precise, which, without getting technical, are created via nonliving, nondiffusive chemical processes), you can determine which specimen derived from living processes.[7]

The carbon-isotope record is gappy—you have to locate ancient marine sedimentary rocks that happen to be on Earth's surface. The oldest such sediments have been identified at Isua, in Greenland, and date from 3.8 billion years ago. Though less striking than the discovery of visually recognizable ancient fossils, this more abstract geochemical evidence is exceptionally important: it tells us that life on Earth has been processing carbon for nearly 4 billion years and that, within the limits imposed by the gappy nature of our ability to find such ancient marine sediments, such life has existed continuously on Earth since the beginning of the Archean.

Green Was Earth on the Trillionth Day

From the discussion of the first two eons, I think you understand that things happen slowly on geological timescales. The same is true of the Proterozoic eon, the eon of basic life. The boundary between the Archean and Proterozoic is drawn some 2.5 billion years ago. The fossil evidence of life in the Proterozoic is very similar to that of the Archean: cellular microfossils embedded within stromatolite-like colonies. But the number and range of fossils is much greater—partly because of changes in the organisms themselves, but also in part because of the fact that there are relatively many more Proterozoic-age rocks conveniently poking out of Earth's surface, patiently awaiting the geologist's hammer (and compared to earlier ones, such rocks tend to have experienced less melting and disturbance from later geological events). Close inspection of individual fossilized cells shows the emergence of the nucleus as a distinct structure—such organisms are today called eukaryotes, a classification that includes all creatures with aspirations to be greater than a bacterium.

Yet something much more fundamental occurred during this eon: oxygen began to appear in our atmosphere. The oxygen appeared as the result of photosynthesis, the process by which life converts CO_2—whether present as a gas in the atmosphere or dissolved in the oceans—into simple sugars to fuel cellular life. Photosynthesis can be written down as a relatively simple chemical reaction in which carbon dioxide, water, and the energy from two solar photons are brought together in a molecule of chlorophyll to produce a sugar (glucose) and molecular oxygen.[8] Though photosynthesis looks like a simple reaction, the details, including the chlorophyll molecule that enables it to occur, are anything but. Therefore, if we view the rise of early life as a progression of increasingly complicated biochemical reactions, then the development of photosynthesis represents a significant number of cumulative evolutionary steps.

So when did photosynthesis evolve? Evolution proceeds through incremental change. It may be that anoxygenic photosynthesis—which does not produce free oxygen as a by-product—evolved first. Anoxygenic photosynthesis is performed by a number of present-day species of bacteria. In these cases, the bacteria employ ferrous iron, sulfides, or molecular hydrogen to react with CO_2 and do not produce free oxygen as a by-product. Were such metabolic pathways evolutionary precursors to oxygenic photosynthesis? We cannot be certain. But some such progression of steps may well have played a role in the evolution of oxygenic photosynthesis.

Evidence for the oxygenation of Earth's atmosphere emerged approximately 2.4 billion years ago, and it was once again geochemical

in nature—this time based on the ratio of sulfur isotopes in Proterozoic rocks. Rather suspiciously, rocks laid down in the few hundred
million years before the rise of atmospheric oxygen show that Earth
was rusting on a planetary scale in what is known as the "great oxidation event." This period in geological history is dominated by features called banded iron formations—swathes of rust red, iron
minerals. Erosion would have washed iron-rich minerals into the
ocean, much as it does today. Once there they would have reacted
with dissolved oxygen—produced by photosynthetic bacteria—and
precipitated out of the ocean to form distinct layers of iron oxide
minerals.

An alternative route to the creation of banded iron formations was
via the more direct oxidation of iron as part of anoxygenic photosynthesis performed by certain species of purple bacteria. Although
we have no evidence of which of these two processes dominated, it
seems reasonable to assume that, at least in part, dissolved iron in
Earth's oceans acted as a sink, or trap, for the feeble trickle of oxygen
produced by early oxygenic photosynthesis. Only once the sink was
full, when the iron had satisfied its corrosive appetite for oxygen,
could the excess escape the oceans and build up in the atmosphere.

Additional clues regarding the metabolism of life's early pioneers
come from the methanogens of present-day Earth: primitive archaea
that convert CO_2 into methane (CH_4) as their source of cellular energy. There is no direct test of ancient atmospheric methane levels
similar to those used to determine ancient oxygen levels. But meth-
anogen activity does provide part of a consistent scenario. Earth's
early atmosphere would have been rich in CO_2 as a result of volcanic
outgassing. Yet importantly, that early atmosphere would have been
devoid of molecular oxygen. Present-day methanogens are exclusively
single-celled prokaryotic archaea, similar in form to the early life captured by fossil evidence. They exist today in low-oxygen environments that are potentially characteristic of Earth's atmosphere before
the evolution of photosynthesis.

We must recognize that "plausible" is not the same as "proven."
Ancient fossils from the Archean yield no direct evidence of their

organisms' detailed metabolic chemistry. We can say one thing with certainty, though: modern methanogens do not like oxygen. As a chemical element, oxygen is just too reactive. It stumbles into a methanogenic metabolism with all the grace of a drunken latecomer to your classy soirée and effectively strangles methanogenic archaea. If early Archean life consisted of methane-producing microbes, then the evolution of photosynthesis and the subsequent rise of planetwide oxygen would have marked one of Earth's first and perhaps greatest extinctions.[9]

But what about our perspective as astrobiologists? The emergence of oxygen in Earth's atmosphere was important because it showed that life had truly become a planet-altering phenomenon—a phenomenon that could be recognized by an astute alien observer of our distant planetary system. In this sense, Earth emerged as an astrobiologically interesting planet some 2.4 billion years ago. To be fair, methanogenic archaea could have so enriched the Archean atmosphere of Earth with high levels of methane that our distant alien astronomer would have taken note a billion years earlier. But lacking a geological test for ancient atmospheric methane, we can only speculate.

As we approach the close of the Proterozoic eon, it is worth noting that the development of photosynthesis arose in part from the evolution of chlorophyll—with its distinctive blue-green tinge. Therefore, the rise of oxygen in Earth's atmosphere would have been accompanied by a greening of the oceans. Earth's surface remained devoid of life. But with the advent of atmospheric molecular oxygen came the added benefit of ozone (O_3). The oceans had perhaps exclusively been the preserve of life until this point: water absorbs ultraviolet solar radiation—both energetic and damaging to life. Therefore, as ozone accumulated in Earth's atmosphere throughout the Proterozoic eon, it shielded the entire planet from harmful solar radiation, exactly as it does today. More subtly, oxygen provided a biochemical turbocharger to the basic metabolic consumption of glucose—at least in those organisms that had the good sense to evolve to exploit it. The rise of oxygen therefore played an important role at the end of the Proterozoic eon in opening up the biggest real estate bonanza—both geographic and evolutionary—that Earth had ever witnessed.

An Evolutionary House Party

The entry into the Phanerozoic eon is one of the more sharply defined geological transitions. Within rock layers created 540 million years ago—and over a relatively short space of geological time—the fossil record erupts during the Cambrian explosion into a riot of new evolutionary forms. The rise to complex life clearly depended on both the specific local environments to which evolution adapted it and the vagaries of chance—consider, for example, the case of the giant meteor impact 65 million years ago, which is believed to have caused the extinction of the dinosaurs (and thus perhaps liberated a place on the evolutionary playing field for advanced mammals). Therefore, attempting to draw general rules for the development of complex life from the specifics of what happened to life on Earth following the Cambrian explosion is likely to be of limited use in our search for life beyond Earth. I don't want to unnecessarily denigrate the Phanerozoic era: it could be argued that the same combination of evolutionary adaptation and chance has affected life throughout its entire history. But it is not unreasonable to suggest that primitive life adapted to the basic planetwide conditions of the early Earth, whereas later, complex life adapted to suit the wide range of specific local conditions present on the modern Earth.

So there is our story of life. It ultimately ends with you sitting there, reading this book. Perhaps now you have a greater appreciation of Darwin's leap of insight when he conquered the depths of time to imagine the origin of life in a warm little pond at the dawn of Earth's history. Somewhere deep within our biochemical makeup resides the memory of that event, and we must ultimately travel back to that time and place if we are to understand the origin of life itself.

The Weight of Numbers

Having taken a broad view of the history of life on Earth, I want to pause for a moment to consider a simple question: suppose you are an alien astrobiologist visiting Earth at some random point in its history to search for life. What kind of organisms would you discover? You would more than likely discover simple microbial life. Think of it this way: bacteria and archaea have existed continuously since the origin of life on Earth. In fact, for nearly three billion years they were the only form of life. Not that I want to look down on higher forms of life, but microbes remain the dominant form of life on Earth today: even if we limit ourselves to single-celled organisms in the world's oceans, there exists some three thousand times more biomass in the form of bacteria and archaea than of humans. Look around you: for each human you see there are three thousand microbial copies of you quietly running Earth's ecosystems. You may sense that I am nailing this point a little too firmly for comfort, but bacteria even defeat the Martian invaders in *The War of the Worlds* after humanity has been laid waste!

Bacteria and archaea are among the hardiest and most adaptable organisms on Earth. They exist in desiccated rocks on the edge of the Antarctic ice cap; they thrive in the superheated water of undersea volcanic vents; they even exist in Earth's depths, abundant in rock samples taken from deep in the crust. These microbes existing on life's limits are loosely described as extremophiles—organisms that thrive in environments intolerable to higher life-forms. Species are grouped by the torments they delight in: thermophiles in volcanic hot springs and undersea vents, halophiles and alkiliphiles in caustic soda lakes, psychrophiles in subzero temperatures. My favorite is *Deinococcus radiodurans*—also known as Conan the Bacterium—a polyextremophile equally at home in low-temperature, acidic, vacuum, and desiccated environments.[10] I talk more about these extremophiles in later chapters—mainly because I want to put our knowledge of them in the context of challenging (yet potentially habitable) environments across the solar system. But for now I sense that you get my point: think of anything you can do biochemically, and there exists a microbe that can do it better.

The Spark of Life?

We have traced the history of life on Earth from the depths of the Archean, 3.8 billion years ago, to the present day. We now have to confront the biggest scientific question concerning life on Earth: How did it begin? How did the chemical environment of the early Earth become a biological one? Should we come up with a satisfactory answer, we can then pose the astrobiological extension of this question: under similar conditions on a distant planet or moon, would life emerge again?

What picture do we have of the proto-Earth in which life arose? How far can we speculate regarding the chemical environment of the Archean? The atmosphere would have been rich in volcanic gases and the vaporized remains of icy comets. Present-day observations of volcanoes and comets point to an atmosphere dominated by carbon dioxide, water, nitrogen, and hydrogen sulfide (among other molecules).

Our modern scientific story resumes in 1924, only some fifty years after Darwin wrote of his warm little pond, with the Russian plant biologist Alexander Oparin. He realized that the only source of oxygen on the early Earth was photosynthesis. He speculated that because photosynthesis was too complex a process to have been developed by the earliest forms of life, the origin of life must have occurred in an environment free of molecular oxygen and its vast, reactive appetite. A few years later, the British biologist John Haldane independently came to a similar conclusion: the early Earth would have possessed an atmosphere largely free of oxygen, and as a consequence a number of simple organic chemical reactions could have created an abundance of more complex molecules that might serve as the precursors for living organisms. Both Oparin and Haldane speculated that the energy to power these reactions would have come from nature—whether in the form of lightning in Earth's turbulent atmosphere or via the ultraviolet sunlight to which the young Earth was exposed.

There the early Earth remained for the next thirty years. A curious young graduate student named Stanley Miller took the speculation of Oparin and Haldane one step further into reality. In 1953, he constructed a wonderfully simple yet highly effective model of the chemistry of the ancient Earth. Miller was guided in his research by his supervisor, Harold Urey, a recipient of the Nobel Prize in Chemistry in 1934 for the discovery of deuterium—an isotope of hydrogen. Taken together, their work is more commonly known as the Miller-Urey experiment.

The experiment consisted of a closed system of glass tubes and a flask of water representing Earth's oceans. Gently heated, the water and its contents evaporated off into the nascent atmosphere— ammonia (NH_3), methane (CH_4), and hydrogen (H_2) in Miller's original experiment. Miller added lightning to his original experiment by using an electrical spark—other versions of the experiment over the years have used ultraviolet light or other sources of energy. A tube from the atmospheric flask led to a simple condenser where gases could condense back into the ocean to begin their journey again. Miller's experiment was a revelation on a number of levels: it is astonishingly simple, yet, amazingly, thirty years passed before an experimenter responded to the ideas of Oparin and Haldane.

But nothing could be as astonishing as the results. Miller left his closed cycle operating for a few days and noted that the initially clear water of the oceans slowly started to discolor from pink to brown. Soon the flask representing the ocean was coated with black tar— clearly his model of the early Earth had been busy! Upon analyzing the contents of the flask, Miller discovered a rich brew of organic chemicals. The most significant result was the discovery of amino acids in the tar-like goo. Amino acids are the basis of proteins, the fundamental language units of our biochemistry.

The Miller-Urey experiment has been copied and varied many times. Rather like cooks experimenting with a classic recipe, scientists have added their own chemical flavors to the experiment's atmosphere: some rich and complex, others spartan and bland. The molecules discovered in these diverse brews included complex sugars and the nucleotide bases present in our DNA.

Whose is the right recipe—the one that best describes the early Earth? Modern ideas concerning the composition of Earth's early atmosphere favor a mix of carbon dioxide and molecular nitrogen over the ammonia-methane recipe originally used by Miller-Urey. Repeating the Miller-Urey experiment with such modern recipes generates much lower yields of amino acids than obtained previously (mainly because molecules such as CO and N_2 are much harder to break apart than CH_4 and NH_3). A second area of active debate concerns the presence of molecular hydrogen in the atmosphere of our proto-Earth. The presence of hydrogen gas would have driven many of the reaction pathways leading to complex organic molecules. But since it is the lightest of gases and easily boils off a planetary atmosphere into space, experts in the field keenly argue about whether it was present (with little hard data to guide them).

In many ways, the exact ingredients in the Miller-Urey experiment are less important than the overall discovery that the conditions on Earth four billion years ago provided a natural pathway for the formation of moderately complex organic molecules. Such molecules play key roles in life today, but it is critical to note that they do not constitute life in their own right. The Miller-Urey experiment and its variations show a series of possible steps en route to life. But the experiment does not tell us *exactly* which reactions were followed on Earth four billion years ago.[11] In this sense, the Miller-Urey experiment is an incredible demonstration that the ideas of Oparin and Haldane were indeed plausible—but tantalizingly, it goes no further. We can demonstrate that our ideas regarding the early Earth are sensible, but we cannot prove that they ever occurred on the road to life.

Importantly though, we have not found any roadblocks on our journey through the world of Miller-Urey: given similar conditions and ingredients on a new world, we can envisage similar results—an environment rich in complex organic molecules teetering on the precipice of life.

It Came from Outer Space!

Would it surprise you to learn that the universe has been performing the Miller-Urey experiment in space for the last thirteen billion
years—that the depths of space are scattered with a dust of organic
molecules similar to those produced in Stanley Miller's flask? You
should be, because it turns out that nature's chemistry set is more
varied and adaptable than we could have imagined.

Some of the best surprises fall from the sky and present us with
small yet irrefutable examples of the universe being cleverer than we
are. One example happened in the small town of Murchison in Australia. A large meteorite fell near the town in 1969, and around one
hundred kilograms of fragments were eventually identified and
studied. The Murchison meteorite is a rocky meteorite—a carbona-
ceous chondrite—an aggregate of loose, partially melted rock. The
surprise contained within was the sprinkling of complex organic
molecules—among them amino acids and nucleotide bases—all of
which display atomic isotope ratios indicative of an origin in space
rather than on Earth. The Murchison meteorite is not rich in such
molecules: their abundance is measured in parts per million at most.

Fired by such discoveries, NASA took the search for the raw ingredients of our solar system one bold step further in 2004, flying
the *Stardust* mission through the tail of comet Wild 2 and returning
samples of the collected dust to Earth. Once again, among the particles of ice and dust captured from their cometary ride were organic
molecules—the simple amino acid glycine once again. We still do not
know for sure how such molecules form—though it is speculated
that the chemical reactions occur on the surface of microscopic
dust grains bathed in ultraviolet starlight.[12] But form they do, in our
solar system and in the depths of dust clouds observed in our Milky
Way galaxy.

Meteoritic debris heavily bombarded Earth during its early
history—which prompts the question: do we even need to resort to
the idea of something like the Miller-Urey experiment occurring on
the young Earth, since the solar system was serving up extra help-

ings of complex organic molecules? Such molecules quite probably arose on Earth due to both processes: exogenous meteoritic delivery and production in situ by a Miller-Urey-type process. Their relative contributions to Earth's organic diet in large part depended on how effective the Miller-Urey process was on Earth: if the appropriate raw ingredients were present, then it seems possible that, in localized spots, the abundance of complex organic chemicals could have been as large as that noted in present-day experiments. An important additional consequence of the Miller-Urey picture is that it brings such molecules together in sufficiently high concentrations that further reactions can occur. And occur they must—for we have not yet crossed the threshold to life. We are tentatively approaching the shadowy frontier between nonliving and living chemistry, and we proceed on uncertain ground.

The Undiscovered Country

A new journey is a lot easier if you have a good map. However, someone has to have traveled before you, taking careful measurements, to finally render them in faithful topographic detail. Does such a map exist for our journey through the murky protoworld of Miller-Urey to the beginning of life on Earth?

Er, no. Our map is at best incomplete. We have some knowledge of our prebiotic starting point—the world of Miller-Urey. We know our present location and surroundings as well as the present-day properties of life on Earth. But we have only sketched out sections of the route between the two, each section a separate stage in the emergence of life. We await a new generation of cartographers who can link them into a definitive understanding of the origin and development of early life.

We paused for reflection earlier in this chapter and defined life as a series of linked physical phenomena. I want to push these ideas further into the realm of the most basic organisms, those that lie on the border between living and nonliving. The molecular chemist Steven Benner has put a simple and effective definition of life forward: life is a self-sustaining chemical system capable of Darwinian evolution. This statement reduces life to the phenomena of order, metabolism, and (flawed) reproduction. Does this view of life at its most fundamental assist us in contemplating its origins?

How can order arise from the apparently random reactions occurring in the world of Miller-Urey? There is more structure in nature than at first meets the eye. For example, the structure of the periodic table derives from the ordered arrangement of protons and electrons within each atom. The relationships between atoms—how many electrons they share and how tight is their grasp—determine the structure of more complex molecules. To take this idea several steps further, just look at how mixing a simple chemical with water can lead to the spontaneous formation of cell-like membranes—in this case, do it while you are washing the dishes, and consider how your detergent combines with water to produce bubbles. Your detergent contains a molecule with a water-loving (hydrophilic) head and a water-hating (hydrophobic) tail. The action of embracing and shunning water leads to the formation of bubbles—which in this case is a pretty good proto-cell.

I am not saying that all Stanley Miller had to do was add some dish soap to his experiment and—hey, presto!—the first living cells would have appeared. But naturally occurring organic molecules—fatty acids in this case—can combine in water to produce cell-like lipid bubbles that fit our ideas of the spontaneous emergence of the first cells—and it is compelling that your own cell membranes are composed of a two-layered lipid (fatty) structure.

When we consider the origin of metabolism, we should recall that it is simply a reaction that breaks down a chemical and releases energy—it just happens to occur in a living organism. From our perspective, we would call the chemical that is broken down breakfast, lunch, and dinner—or at a simpler level, glucose. Such simple sugars can arise in the world of Miller-Urey and can be spontaneously broken down to release energy. If, at some point, such a reaction were incorporated into a cell, you would end up with an energy-producing cell—albeit one that will get hungry and soon require more fuel. This is only one example among many that could be suitable candidates for the first metabolic reaction.

But what would cause these reactions to persist when separated from the coincidence of local conditions that initially favored them? Both a recipe and an agent are needed to drive them. Within modern cells, such recipes are encoded in the sequence of genes that constitute our DNA. But DNA is complex. Its biochemical cousin ribonucleic acid (RNA) is simpler, yet retains the dual role of information carrier and reproducer. We don't know whether the first organisms were based on RNA or on some simpler, more primitive ancestor.

Richard Dawkins and others have imagined the emergence of the "replicator," a simple organic molecule—perhaps life's simplest level—with the single yet unique property of being able to collect fragments of other molecules to its own structure and arrange them into a copy of itself. We can further speculate whether the emergence of an organic molecule of such specialized properties might have required a nonbiological scaffold or a boost onto the ladder of life. Did the repetitive crystal structure of wet clays or iron pyrite fortuitously provide a mineralogical backbone that allowed organic structures adhering to it to take on the properties of the replicator?

We can gaze out onto the undiscovered country that contains the origin of life and see the distant peaks of order, metabolism, and reproduction. We can return to our laboratories and sketch out their biochemical contours. But essentially all experiments that we can perform will be demonstrations of plausibility rather than exercises in proof. We may eventually demonstrate that a series of specific physical reactions will lead to the formation of primitive organisms similar to our ideas of what early life must have been like. Yet we must remember that our ideas of early life are based on the evidence of later, complex life, and there is more than one way to skin a cat. So it is not clear how science will proceed from having a series of plausible ideas regarding the emergence of life to determining exactly which microbiological path our earliest ancestors took.

A Second Genesis?

Did life arise on Earth more than once? If so, could it have been an independent strain of life with a distinct biochemical identity? Is there any evidence in the fossil record of such organisms? Might they even exist on Earth today—a hidden, shadow biosphere? While this may seem like a question more appropriate to a book focusing on life on Earth than to one aiming to search for life beyond Earth, it would be an embarrassing oversight to cast our search for alien life to the distant worlds of space while missing it under our feet.

So first, did life arise on Earth more than once? At present we have no evidence that it did. Had it done so, we have to accept that the signature of a separate form of life could exist in the fossil record. But it might be a challenge to recognize it. Claims that what appears to be ancient fossilized life formed part of our family tree are subjected to considerable scientific scrutiny and are largely accepted or rejected based on cellular comparisons with present-day life. But how do you discover a completely new branch of ancient life in the fossil record when all our techniques for confirming a sample are based on shared features with existing life? Although such paleobiological needles may exist in the haystack of Earth's fossil record, the frustrating reality is that we may well lack the ability to recognize them.

Could such a separate strain of life exist on Earth today? Once again, I have to give an emphatic yes: as long as there remain vast unexplored sections of our planet—and I am thinking here of the geochemical landscape of Earth's crust—there is the potential to discover isolated (yet potentially thriving) habitats exploited by new forms of life independent of our own.

Beyond this remains the interesting—yet essentially unanswerable—question of why, if we imagine independent strains of primitive life arising at multiple locations and multiple times on the early Earth, each one perhaps a lucky roll of the molecular dice, did only our strain survive and persist until today?[13]

An Astrobiological Road Map

We began this chapter with a pair of linked questions: what on earth is life, and what is life on Earth? We have covered a lot of ground, but we can end this chapter with some answers and ideas.

Life is a series of linked phenomena that we have simplified at the biochemical level to those of order, metabolism, and reproduction. As astrobiologists, we would do well to consider searching for primitive life similar to the bacteria and archaea that have dominated the history of life on Earth and dominate the mass of living organisms today. Our search for life will be based in large part on identifying biochemical reactions and cycles—and on excluding exotic yet distinctly nonbiological reactions that might masquerade as life. We may have to follow the example of fossil hunters delving deep into Archean rocks for evidence of early life on Earth and focus on the hunt for microscopic cellular order. I think we also understand that there is no single rule book for identifying alien life. If we imagine a physical sample painstakingly obtained from an alien environment by a remote probe or a human scientist, each new rock or scoopful of soil will present its own unique challenge to those who seek life within it.

Life on Earth arose from a series of complex natural chemical reactions promoted by the conditions present on our young planet— liquid water, organic chemicals, and energy. It also arose early in Earth's history—effectively as soon as surface conditions permitted, at the end of the Hadean. Though our knowledge of life's origins remains incomplete, the gaps are not so wide as to be insurmountable by leaps of science, as opposed to leaps of faith. Like Darwin, we can imagine a warm little pond on another planet or moon—in our solar system or beyond—where life emerged from similar principles. We have seen in our own history how microbial life emerged to become a planetwide, planet-altering phenomenon and created a chemically imbalanced atmosphere rich in oxygen—an atmospheric biomarker visible to distant observers. We will discover in later chapters how close we are to becoming that distant observer, searching for the signature of life in extrasolar planets.

Finally, we also asked how much of this should matter to an astrobiologist seeking alien life in alien environments. We have learned something of the principles of life rather than the particulars of life on Earth. Our search for life should not be excessively Earth-centric, but we must make informed choices about where to focus our search. It is therefore time to leave Earth behind—perhaps gratefully so after I have packed our bags so full—and to explore the solar system for the first time as astrobiologists seeking new habitats for life.

One of the biggest shifts that astrobiology has brought to our view of the solar system has been to consider it a habitat capable of sustaining life. At present we know of life only on Earth, the third planet from the sun. But we have learned that the conditions that led to the emergence of life on Earth are not as special as we once thought. The early Earth provided three fundamental ingredients: energy, liquid water, and complex organic chemicals. A fourth feature that has aided life on Earth is the stability of this environment: the presence of the above three ingredients has remained more or less stable over the eons of Earth's history, permitting life not just to arise but also to develop and flourish.

We shall discover on our tour of the solar system that these conditions are not unique to Earth: energy is present (though sometimes scarce) even at the farthest reaches of the sun's influence. Liquid water has been hinted at (really quite strong hints, actually) at several important locations, and real liquid—in this case, ethane and methane—has been observed directly on Saturn's moon Titan. And finally, complex organic chemicals have been detected throughout the solar system—on the moons of Jupiter and Saturn, and hitching rides on comets and asteroids.

So should we expect life to be abundant in the solar system—awaiting our arrival with barely suppressed alien glee? Well, that's really the $64,000 question, isn't it?[1] We still don't know—even on Earth—how nature proceeds from a list of enticing ingredients to life itself. Our present-day sense of anticipation that life awaits us in the solar system is based on detailed knowledge of the physical conditions of its planets and moons. In addition, with the discovery of extremophiles on Earth thriving at the very edge of conditions previously thought to be lethal, we have learned, and continue to learn

53

even today, that life is much more durable and adaptable than previously thought.

In this chapter I want to set the scene for our later detailed exploration of the planets and moons of the solar system: What is the physical makeup of our solar system—the geography, if you like? Which planets and moons might be of particular interest in our search for life? Which, if any, can we discard as unlikely habitats? We will take some time to appreciate how nearly sixty years of space exploration has allowed us to get up close and personal with the moons and planets of the solar system: What have we learned from past space missions? How do they reach their destinations? How are they built? Who pays for them? All of this will, I hope, prepare us to make a leap into the solar system and to consider future missions to our top planetary and lunar targets—and to gauge their prospects of success.

The Sun Is a Mass of Incandescent Gas

In many ways, the sun *is* our solar system. The simplest way to appreciate this is to take a sheet of graph paper. Draw a rectangle 10 squares by 100. Of the total 1,000 squares, almost 999 are required to represent the mass of the sun. The remaining square and a bit represents, mostly, the mass of Jupiter and Saturn. We merit barely even a smudge.

With this drawing, it is easy to visualize the extent to which the sun dominates the solar system. Nuclear fusion in the core of the sun releases energy in the form of photons plus their more elusive cousins, the neutrinos. These photons are incredibly energetic—we would call them gamma rays and X-rays at this point. As they diffuse outward, absorbed and emitted by the atoms in the sun's outer atmosphere, they decrease in energy until they escape the haze of the sun's photosphere as the sunlight you observe today.

Solar energy powers almost all life on Earth today: photosynthetic organisms and everything higher up the food chain that feeds on them. The exceptions to this rule are some of the most astrobiologically interesting examples of life on this planet: extremophilic mi-

54

crobes living near undersea volcanic vents and iron-oxidizing bacteria deep within Earth's crust. These guys merit their own section—so hold that thought!

A Tour through the Burbs

Stepping out from the sun, we enter the realm of the planets. Looking back to the age when ours was the only planetary system we knew of, we were all rather pleased with it. It seemed so ordered, just how our theories suggested it should be. We will learn later how the discovery of new planetary systems around distant stars has turned upside down many of our cozy ideas of how such systems should look. But for now I want to cast a glance over our own solar system.

Closest to the sun, in the well-warmed inner regions, come the terrestrial—or Earth-like—planets: Mercury, Venus, Earth, and Mars. Though they vary in size and surface conditions, each is basically a lump of rock (iron and silicate minerals). There are few moons among the terrestrial planets: our moon, plus Deimos and Phobos orbiting Mars. Beyond Mars lies the asteroid belt, a shattered relic of the early solar system that never escaped from Jupiter's gravitational shadow to coalesce into a planet in its own right.

En route to Jupiter we pass a subtle yet important boundary—the frost line. At this point, the waning light of the sun weakens to the point that simple gases such as water, ammonia, and methane can condense to form solid ice grains. As a result, beyond the frost line, not just rock but ice as well can join in the collisional growth of planets—in this case forming the gas giant (or Jovian—Jupiter-like) planets that dominate the outer solar system. Jupiter comes first, followed by Saturn, Uranus, and Neptune.[2]

In contrast to the inner solar system, the outer planets possess moons in abundance: Jupiter has some 67 moons, and Saturn has over 150. The largest of these moons rival the size of Mercury or our own large moon. They are worlds in their own right, ripe for exploration. This disparity in the abundance of moons between the Jovian and terrestrial worlds is simply a matter of mass. In the swirling disk of gas and dust that made up the nascent solar system, the gas giants grew to become large and massive. They accreted their own miniature disks of gas and rock, and from each of these grew their abundant retinue of moons.

As we head past Neptune, the sun appears less than one-thousandth as bright as it does from Earth. We have entered the dark realm of Pluto—first glimpsed by Clyde Tombaugh in 1930. In the late 1990s and early 2000s, a new generation of large modern telescopes scoured the outer solar system and found more Pluto-like chunks of rock: some bigger, some smaller, all part of the outer-solar-system debris disk we call the Kuiper belt. Pluto no longer stood out from the crowd—either all of these rocks were planets or none were. In 2006, the International Astronomical Union decided that none of them were: they, along with the larger asteroids, would henceforth be known as dwarf planets, and unless our current view of the solar system changes, there they will remain.

The scale of the solar system is most easily appreciated by considering the time it takes for a photon of light to travel from the surface of the sun to each planet. It takes nearly 8 minutes for a photon to reach Earth. The sun you see is an image of what it looked like some 8 minutes ago. The sun of the present moment is hidden from us behind a curtain of time that the finite velocity of light cannot penetrate. It takes each photon a further 4 minutes to get from Earth to Mars. If you think about it, a radio signal is simply a stream of low-energy photons, so it takes a radio message or TV signal 8 minutes to make the round trip from Earth to Mars. This helps explain why Mars rovers are moved using short sequences of simple commands rather than driven via a joystick with an 8-minute delay (you would get stuck or crash before you knew it). Jupiter is 42 minutes from the sun as the photon flies, and Neptune, the last of the gas giants, is four hours away. If we consider Pluto our marker of the outer limits of the solar system, it takes a photon 5 hours and 20 minutes to complete its journey to the dark depths on its way out into the Kuiper belt.

You should now have a clearer sense of your place in the solar system and its scale. Your only remaining task is to watch the opening scene of the movie *Contact* and to criticize it accordingly!

The Long Reach of Sunlight

We have noted that the sun is the powerhouse of the solar system and supports almost all life on Earth. But how far does the life-giving reach of the sun extend? When does faint become too faint for life?

Earth's upper atmosphere is bathed in approximately 1,370 watts of solar energy per square meter.[3] The sum of this energy—received day in, day out—drives almost all life on Earth and every last bit of weather. The amount of sunlight available to each planet and moon in the solar system can be thought of as the base budget for life—at least life, like photosynthetic bacteria, that converts solar energy into usable fuels.

So how much sunlight is available? Mercury orbits close to the sun and receives some six times as much solar energy as Earth. Mars, farther from the sun than Earth, bathes in only about 40 percent of the solar energy received on Earth. As we move into the outer solar system the sun's influence declines precipitously: Jupiter receives only 3 percent of one Earth dose of sunlight. At the frigid radius of Pluto, it is less than 0.1 percent.

Perhaps the more interesting question is how much light does sun-loving life need in order to survive? Once again, our experience on Earth has been crucial in revealing the surprising limits of life. Photosynthetic bacteria have been identified up to one hundred meters deep in the waters of the Black Sea. Their metabolism, though, is based on anoxic—or oxygen-free—photosynthesis, which produces sulfur compounds rather than molecular oxygen. Such bacteria may represent living relics of the first photosynthetic organisms. At such depths, only 0.05 percent of light is transmitted from the surface—light levels so faint they mimic the solar illumination of Pluto. The low light levels nonetheless remain biologically useful, with each bacterium carefully harvesting a photon every few hours and using its tiny flash of energy to keep its metabolism rolling.

So when we look out into the solar system, there does not appear to be any point or boundary beyond which we can state definitively that the sun is too weak to support photosynthetic life. The reach of the sun stretches, at least weakly, into the farthest reaches of the solar system and provides a source of energy for life—should any exist to use it.

Liquid Refreshment

So there is plenty of sunlight throughout the solar system, and we have already determined that simple organic chemicals are also present. What about water or—in order to keep an open mind—liquid? This chapter is all about focusing our attention—and limited resources—on the most promising habitats for life in the solar system. So it is time to make some progress and get nasty.

Mercury—you have no atmosphere, you are blasted by the full force of the solar wind, and your daytime surface temperature is 700 Kelvin—you are out. Venus—the dense atmosphere does not put me off, but your surface temperature is hotter than Mercury's (737 Kelvin). Although life may exist in a form different from that on Earth, the proteins that run our biochemistry break down above 400 Kelvin. You are basically a dry, superhot oven. The moon—no. No atmosphere, no liquid water. Very nice to visit for a few days, but we are headed elsewhere. The Jovian planets? They call you gas giants for a reason. In 1995, the *Galileo* spacecraft launched a probe into Jupiter's atmosphere—plunging 156 kilometers through the clouds before the rising temperature killed the onboard systems. Atmospheric models of Jupiter and the outer gas giants include the existence of exotic liquid layers. But it is difficult to speculate about what type of life they might host, and they are very difficult to get to (the probe launched by *Galileo* died well before this point). Pluto and the Kuiper belt? Way too far, and no hope of liquid when we get there.

Phew. I am sorry if the above upset you, but it had to be done. Am I ruling out the existence of life on the planets and moons that I treated so harshly? Not at all. Are they even close to the top of my priority list for future solar system life-hunting missions? You guessed it. What are we left with? The table of contents leaves no room for surprises regarding the idea that we will spend a lot of time (and expend a lot of words) considering the possibility of discovering life on Mars, Jupiter's moon Europa, or Saturn's moons Enceladus and Titan.

My decision to focus on these potential habitats at the expense of other locations in the solar system is based in large part on what we have learned in previous chapters about the requirements of life on Earth. We will see that Mars, Europa, Enceladus, and Titan present not a cast-iron case for believing life is present there, but rather enough fragments of evidence for the combination of liquid, organics, energy, and stability that push them to the top of our wish list for solar system exploration. Now we are getting somewhere.

Goldilocks and the Three Planets

In the story of Goldilocks and the three bears, we learn about a little girl's search for the porridge, chair, and bed that are "just right" for her during her uninvited visit to the traumatized bear family. Strangely enough, the Goldilocks principle of searching for planetary conditions that are "just right" for life has been applied in astrobiology as well. In this case, the idea is known as the "habitable zone": the range of orbital distances from a star at which the surface temperature of a planet is "just right" for life, in this case between the freezing (273 Kelvin) and boiling (373 Kelvin) points of water. I don't need to repeat my usual warning about making our search for life too Earth-centric for you to realize that the idea of the habitable zone must handled with care.

The main reason for caution centers on the uncertain properties of planetary atmospheres. First and foremost, you have to have one. Earth lies smack in the middle of the solar system's habitable zone, yet take away the surface pressure created by our atmosphere, and our planet's water would boil away into space. Second, while the temperature of the parent star and the orbital distance of the planet are the main factors in determining a planet's surface temperature, any atmosphere that a planet may have adds a considerable range of uncertainty regarding the actual surface temperature.

Take three of the terrestrial planets of our solar system as an example: Venus, Earth, and Mars. For many years they formed the basis for comparative planetology—the attempt to understand how small changes in a planet's bulk properties (mass, rotation speed, orbital radius) would result in considerable differences in the surface conditions on each planet. They also lie close to the approximate boundaries of the solar system's habitable zone, and they serve as a warning of the havoc that variations in planetary properties can wreak on a simple idea.

Venus is very similar to Earth—it is about four-fifths as massive as Earth (0.8 Earth masses) and orbits slightly closer to the sun (about 0.7 times the Earth-sun distance). The theoretical temperature[4] that Venus would have in absence of an atmosphere is about 260 Kelvin (about −13° Celsius). But Venus has a superdense, CO_2-rich atmosphere that creates a spectacular planetary greenhouse effect and results in a surface temperature of 737 Kelvin. Any liquid water that may have existed on Venus has long since boiled off into the atmosphere (where it contributes to the greenhouse effect), and even water trapped in the mineral structure of surface rocks would have been baked away into vapor.

Mars is smaller than Earth—only one-tenth of its mass—and orbits one and a half times farther from the sun. It has a thin CO_2 atmosphere that creates a surface pressure only 1 percent of Earth's. The blackbody temperature of Mars is 210 Kelvin, and the measured surface temperature is only a few degrees warmer—the result of an exceptionally weak greenhouse effect. The frigid temperatures have locked away almost all the water and carbon dioxide on Mars—in polar ice caps and a subsoil ice layer stretching over much of the planet.

It is fun to play the imaginary game of shuffling the planetary pack and guessing how swapping the positions of the planets would change their properties. What would happen if we swapped Mars and Venus? Mars is easy, I think. If the atmosphere stayed as it is today, then the surface temperature of Mars would be pretty much 316 Kelvin, or 43° Celsius. We can speculate whether the higher surface temperatures at its shuffled location would melt the ice caps (CO_2 and water), creating an abundant yet short-lived atmosphere (remember that Mars has no magnetic field and no volcanism).

Venus is a bit more difficult to predict: its atmosphere generates 400 degrees of additional warmth above its theoretical temperature. Put it at the orbital radius of Mars and it would still have a surface temperature of over 600 Kelvin (if the atmosphere did not collapse under its own weight). Part of the reason why Venus is so hot today is because it experienced what we assume was a runaway greenhouse effect early in its history. Venus became too hot, too early. Its CO_2 atmosphere was mixed with abundant water vapor as any oceans present boiled away. This in turn strengthened the greenhouse effect of the atmosphere in a catastrophic feedback loop.

If Venus existed at the orbital radius of Mars, would this runaway effect still have occurred? The short answer is that we don't know—despite the fact that this question has provided both entertainment and frustration in equal measure for those who investigate planetary atmospheres in vast computer simulations.

So be wary of drawing simple conclusions from a press release or news article telling you that a newly discovered planet lies within the habitable zone of its parent star. In the absence of hard data on the atmospheric composition, any surface temperature you calculate will be speculative (and if you don't include an atmosphere, then why would you expect liquid water to be present?). Just like Goldilocks, without better data we cannot tell whether the porridge (or in this case the planet) will be just right for life.

Panspermia: The Theory That Dares Not Speak Its Name

At the beginning of this chapter, I suggested that we consider the solar system to be a habitat capable of supporting life. But was that suggestion a little disingenuous? The habitats that we have spoken of so far have been located on Earth. Each forest, pond, river, and plain is linked within the single, overarching habitat that is our planet. Species can move between habitats, sometimes freely, sometimes with difficulty, yet migration is always possible. When we extend the idea of habitats to the solar system, does each planet and moon represent a true island in space, or is interplanetary migration possible? Put another way, could a species of primitive life arise on one planet or moon and then migrate to another via natural means?

The idea that life may behave in this manner is known as panspermia.[5] In the simplest scenario, a primitive organism living quietly on given planet is rudely blasted into space by a meteorite impact and becomes a floating chunk of space debris itself. Our intrepid microbial astronaut may then enjoy a few million years floating aboard its chunk of rock before falling to the surface of a different planet or moon as a meteor in its own right. We know that the above happens in our solar system: a small fraction of meteorites discovered on Earth were blasted off the surface of Mars and the moon before falling to Earth. The only missing component is the presence of primitive life in such meteorites.[6]

Quite frankly, the idea sounds nuts, but that doesn't mean it couldn't happen. There would of course be challenges along the way: our bacterial astronauts would have to survive the blast of the initial impact that hurled them into space. Then would come the long, long period of dormancy, floating through the void between planets— exposed to the vacuum of space and streams of ionizing radiation— and when I say long, I mean potentially millions and millions of years. Finally would come the fiery, meteoritic descent to the surface of the new planet and the juddering impact of arrival.

The lesson from science is that you should not rule anything out until you have tried it, and to their credit, a number of scientists have attempted to re-create part of the journey that primitive life would have made as it traveled around the solar system. Many species of bacteria, archaea, fungi, and lichens have gained their astronaut's wings during flights on single rockets, the space shuttle, and the International Space Station (ISS)—in particular, the ISS EXPOSE facility, which flew on the front of the *Columbus* science module for eighteen months during 2008 and 2009. The facility was a test bed for exposing various samples—both living and nonliving—to the environment of space to see how they would fare.

Are the vacuum, cosmic rays, and temperature extremes of space lethal to primitive life? The answer is a pretty definite no. Across many samples and species, primitive Earth life has shown a remarkable ability to shut down, shut up, and survive. Growth and metabolic activity cease completely. Many cells die, many are damaged, but there are invariably survivors able to overcome the harsh realities of space.

What does all this mean for panspermia? Once again, we are defeated by the idea of deep time: flying some rock-dwelling fungus on the space station for eighteen months is a very interesting experiment, but it tells us almost nothing about the ability of organisms to survive an interplanetary journey of a million years. The idea of an organism existing in a near-death state for that amount of time really does seem implausible. If we ever discover an organism that can grow and metabolize in space—or at least do so while hidden deep within a rock floating in space—then that may cause a lot of rightly skeptical scientists to think again. In the meantime, my hope is that somebody persuades a space agency to drop a microbe-laden rock from orbit back to the surface of Earth and check for survivors!

To Boldly Go . . .

We have outlined the geography of the solar system and our top targets for life, so it is now high time to go visit. The search for life in the solar system is both unique and exciting because we can physically travel to the habitats of interest, perform science in situ, and return samples to Earth for more extensive study. Space missions to explore the solar system come in all shapes and sizes. The hardest part is simply getting off the ground: the bulk of any spacecraft—the rocket and explosive cargo of fuel—gets you from Earth's surface to a high-flying orbit. Once there, the freedom of space is yours: relatively little fuel is required to navigate through the solar system, especially if you are sensible enough to catch the gravitational assistance of a passing planet. Despite the varied science that each space mission performs, all of them essentially fall into one of four main types: flyby, orbiter, lander, and sample return.

Flyby: The Briefest of Planetary Encounters

A flyby mission does just that—it approaches a planet, often at high velocity, turns on its cameras and instruments, and records as much data as possible during the blurred planetary passage. The *Pioneer 10* mission to Jupiter is a classic example of a flyby mission. Launched in 1972, it became the first spacecraft to traverse the asteroid belt. Taking some twenty months to approach the Jupiter system, *Pioneer 10* homed in on its target at a blistering 130,000 kilometers per hour. Let me tell you—that is fast.

The critical mission segment began at 12:26 on December 3, 1973, when *Pioneer 10* flew past the moon Callisto and entered the inner Jovian system. *Pioneer*'s suite of cameras and remote sensing instruments were a blur of activity while capturing every possible view of Jupiter and its large moons as it raced by. This brief yet close encounter lasted for sixteen hours before *Pioneer* shot out of Jupiter's shadow and flew off into the darkness of the outer solar system.

After that one frantic day, the main mission was effectively over. What did we learn from our few hours of Jupiter flyby? Undoubtedly, the most important discovery concerned Jupiter's vast magnetic field, ten times stronger than Earth's. More subtly, we saw—with a suite of advanced cameras and sensors—Jupiter's layers upon layers of dynamic clouds, the awesome scale of the Great Red Spot, and our first fleeting glimpses of Jupiter's large moons: Io, Europa, Ganymede, and Callisto.

Our moon in a sense became a world with Galileo's observations of 1609, and in the same way, Jupiter and its moons became worlds on December 3, 1973. *Pioneer 10* continued to broadcast to Earth until January 3, 2003, when the fading power of its radioisotope thermoelectric generator—a nuclear battery, if you like—could no longer operate the radio that served as its link with Earth. *Pioneer 10* brought us for the first time up close and personal with Jupiter; *Pioneer 11* went one better: encountering both Jupiter and Saturn in a whirlwind tour of the solar system's largest planets. Each of these missions was but a precursor to NASA's grand tour missions of *Voyager 1* and *Voyager 2*—which I celebrate later in a section of their own.

Orbiter: To Catch a Passing Planet

An orbiting spacecraft performs many of the same maneuvers that a flyby mission does, with one important exception: it brakes. It must carry enough propellant to decelerate and steer the spacecraft into a stable orbit around the planet or moon of interest. Once there, only a small amount of additional thrust is required to periodically correct or change the orbit. For the time spent studying a single environment, entering orbit makes a big difference. Much as Earth-orbiting satellites map and study the surface of our planet at their leisure, an orbiting spacecraft around a distant planet can spend an extended period studying the whole planetary surface—even revisiting large sections to look for changes over time.

Mariner 9 became the first spacecraft to orbit another planet when it arrived at Mars in 1971. One of its later cousins, the *Mars Reconnaissance Orbiter (MRO)*, showcased the breathtaking views one can obtain from nearly three hundred kilometers above the Martian surface. The *MRO* is basically a spy satellite; it performs the same tasks on Mars that a host of satellites—both civilian and military—perform on Earth. The High Resolution Imaging Science Experiment, or HiRISE, is the heart of the mission: a $40 million digital camera equipped with a half-meter aperture zoom lens. Each image comprises three to five gigabytes of data and can capture features on the Martian surface as small as one meter across.

The primary objective of the *MRO* is to explore the surface of Mars and to help plan future missions—an endeavor it has performed since 2006, with few signs of slowing up. Its high-resolution images were critical in planning the drop zone for the *Curiosity* rover mission. In addition, the *MRO* has notched up a number of science highlights of its own: capturing the aftermath of a meteorite impact on Mars that excavated freshwater ice from the subsoil; catching rock avalanches in the act and seeing plumes of dust and rock crashing down the slopes; and perhaps most enigmatic and important of all, witnessing what appears to be seasonal melting of subsurface ice during each Martian spring.[7]

The other important fact about the *MRO* is that all its images—and those of many other Mars-orbiting missions—are public and awaiting your careful eye. In the laudable tradition of citizen science, NASA has created the "Be a Martian!" website to welcome the public and enlist its help in studying the huge amounts of imaging data gathered across the expansive surface of Mars. The aim is to build a single comprehensive map of the surface. So go for it, get lucky, and find some fresh tracks!

Lander: One Small Step

Okay—so you are in orbit about a distant planet or moon. Having traveled such vast distances to get there, the surface appears tantalizingly close. But beware—think of how much energy you had to expend to get from Earth's surface into orbit. In descending to the surface of another planet or moon, you will gain a similar amount of energy in return. We need a controlled descent. And because of the long time required for a signal to travel to Earth and back, we need to design one that requires no human intervention: we must plan everything and leave a well-programmed computer to make the critical decisions.

A lander will usually be carried as an additional component of an orbiting mission. For example, NASA's *Cassini* mission to Saturn carried the European Space Agency's Huygens lander along for the ride. It ultimately made a safe and (literally) groundbreaking descent to the surface of Titan in 2005. Should it survive the descent to the surface, a lander will provide the ultimate "ground truth"—a record of the exact conditions on the surface and how they change with time. It can scoop up and analyze the surface soil, taste the air, and show close-up images from the surface of a new world.

Physical measurements from the surface of a new world can bring unexpected discoveries. When the dual *Viking* landers touched down on Mars in 1976, they sampled the atmosphere for the first time and recognized a faint, familiar scent. Gases of a similar isotopic signature were known from a rare class of meteorites discovered on Earth.[8] When two and two were put together in the mid-1980s, it became clear that these meteorites were of Martian origin—blasted from the surface of Mars, they later fell to Earth as meteorites themselves (think of this as a no-cost sample return mission—more later).

Rovers: A Wandering Scientist I

A rover is basically a mobile lander—performing the same kind of science on a moving platform that can access a wider range of terrain. Many of us are familiar with the Mars rovers that have explored Mars successfully since *Sojourner* landed on Mars in 1997. But I want to give an honorable mention to two truly pathfinding rover missions from the early 1970s: *Lunokhod 1* and *Lunokhod 2* were Soviet rovers that explored the surface of the moon in 1970 and 1973.[9] Lost in the glare of publicity given to the U.S. manned exploration program, these two rovers dutifully explored the lunar surface, testing the soil mechanics, measuring the local magnetic field, and recording the local "space weather" of the solar wind.

Powered by a combination of solar arrays and radioisotope thermoelectric generators, the rovers pioneered many of the design innovations familiar on later rovers: a multiwheeled chassis, a cluster of forward-facing instruments, and elevated navigation cameras. Since a signal takes less than three seconds to travel between Earth and the moon, the *Lunokhod* rovers were controlled directly from Earth, unlike their later Martian cousins. And these Soviet-era rovers can hold their navigation cameras high: only in mid-2014 did NASA's Mars *Opportunity* rover finally overtake *Lunokhod 2* to become the farthest-traveled solar system rover, with just over forty kilometers on the odometer (compared with thirty-nine kilometers for *Lunokhod 2*).

Sample Return: Leave Only Wheel Tracks, Take Only Rocks

Finally, a sample return mission performs all the above mission activities and then packs either a small or large amount of surface material (rocks, soil, cockroaches) into a return unit that blasts off from the surface, reunites with the orbiter, and uses a nice full tank of propellant on the long journey home.

Why go to all that extra effort? A remote mission—whether by lander or rover—carries with it a suite of experiments designed to learn about local surface conditions. But the onboard experiments are very limited compared with what could be achieved in an Earth-based lab (bigger, better equipment, more sensitive tests). Perhaps the most important factor missing from a remote mission is adaptability: once your experiment package is fixed, that is it. You can't perform new experiments based on what you learned from the first pass—you cannot dig deeper into the sample and learn more.

The most ambitious sample return mission planned for many years was the Russian Phobos-Grunt mission, which flew in November 2011. It was an audacious attempt to obtain a physical sample of Mars by landing not on the planet itself, but on its small moon Phobos. The idea was that it is easier (and requires less propellant) to get off a moon than a planet. Its planned sample of two hundred grams of Phobos soil would have been the first large sample of extraterrestrial material returned to Earth since the lunar missions of the 1970s. Unfortunately, Phobos-Grunt returned a little earlier than expected: once in low-Earth orbit, the spacecraft systems failed, and it floated helplessly for a few weeks before atmospheric drag initiated a fiery and destructive return to Earth.

Footprints in the Regolith

Where is my special section extolling the advantages of human spaceflight? Sorry to disappoint you, but I include human spaceflight with sample return missions because they perform the same tasks— land, perform science, plant a flag, load up with rocks, and return home. The only physical difference is that a sample return mission does not have to split its priorities by keeping a group of passengers comfortable for the duration of their journey. Don't get me wrong—I think that human exploration of the solar system is exceptionally important, both to chart a path for the future of human development and to fire the imagination of a planetful of budding scientists and engineers. But when contemplating the search for life in the solar system, I am not interested in going out there and shaking hands with it. Compared with the risks and expense of human spaceflight, we can potentially learn a greater amount, about a wider range of planets and moons, at a fraction of the cost, by directing our efforts toward autonomous robotic probes and audacious sample return missions.

Brass Tacks

I have talked before about the need to direct our limited resources toward the most promising potential habitats in the solar system. So just how limited are those resources—and by resources, I mean money: your money, my money, taxpayers' money.

How much does a major interplanetary mission cost? The twin *Viking* missions to Mars in 1976 cost a total of about $1 billion (approximately $4 billion today). That bought you two orbiting spacecraft, each equipped with a state-of-the-art lander featuring a suite of scientific instruments. The Mars Science Laboratory, currently operating on Mars—also known as the *Curiosity* rover—has a total mission budget of $2.5 billion. That buys you a semiautonomous rover the size of smart car, powered by a nuclear battery. Looking to the future, the ESA is currently developing the *JUICE* (*Jupiter Icy Moon Explorer*) mission to orbit the largest moons of Jupiter—Europa, Ganymede, and Callisto—with a price tag of 900 million euros.

So if I gave you $4 billion and told you to go and look for life in the solar system, how would you go about it? Would you place all your chips on one ubermission to explore a single location in depth? Would you hedge and send two large missions (about $1 billion each) to your two best candidates—leaving $2 billion to fund five smaller missions elsewhere to make sure you had not missed anything? Or would you assume that you did not know enough about the solar system and spend your money sending small, cheap missions to learn basic facts about the planets and moons of the solar system before committing to a larger project?

Don't worry if you find it difficult to come up with an answer—so do the national space agencies whose job it is to make such plans. Numerous calls for ideas are issued as part of their ongoing mission to explore beyond Earth. Teams of scientists respond with detailed mission plans designed to answer clear, physical questions. These ideas are then pitted against one another in (generally) winner-take-all competitions before the victor starts on the long road to construct, launch, and fly its mission.

NASA's *Cassini* mission to Saturn and its moons remains in orbit today, returning new scientific observations and many new surprises. Among *Cassini*'s many achievements, two of the most exciting have been the discovery on Titan of lakes of liquid ethane and methane, together with the revelation that plumes of water vapor and ice crystals are being ejected from the tiny ice moon Enceladus.

Cassini arrived at Saturn in 2005 after an eight-year journey. Its route was circuitous: it flew first to Venus in a slingshot trajectory that took it back to Earth and then past Jupiter before the long cruise to Saturn. Upon reaching Saturn and its largest moon, Titan, it dispatched the ESA's *Huygens* lander to the surface of Titan—a memorable achievement that marked the first landing in the outer solar system. It should be noted that the first meetings between NASA and the ESA to plan this impressive joint venture began in 1982. From there it took twenty-three years to get to the destination, and we are still reaping the scientific benefits some ten years later.[10]

Do you remember back in chapter 1 when I asked you when you thought that alien life might be discovered? If you answered that it would be in one hundred years, may I politely suggest that you start thinking about your mission and preparing your proposal?

Voyager 2: *The Grandest Tour*

Most space missions reveal the secrets of a single moon or planet. Well-planned ones like *Pioneer 11* visit two—in this case, Jupiter and Saturn. But one mission stands out for having soared through the solar system to fly by each of the giant planets—Jupiter, Saturn, Uranus, and Neptune—and their accompanying moons.

The *Voyager* missions were conceived in 1964 by Gary Flandro, a young graduate student using a slide rule. The idea of the gravitational slingshot, arcing a probe around a planet to gain a dramatic boost in velocity, had been applied to some of the earliest lunar probes. During a summer student placement at NASA's Jet Propulsion Laboratory, Flandro saw that the planets would align in the late 1970s and 1980s in way they might not repeat for over a century. The opportunity existed to send a series of probes to visit the giant planets of the outer solar system—the velocity required to reach the next planet out would be gained from the gravitational encounter with each planet in turn.

The possibility fired the imagination of NASA and led to a series of probes that made up what became known as the Planetary Grand Tour program. *Pioneer 10* tested the spacecraft design on a single-planet ticket to Jupiter in 1973. *Pioneer 11* took the same spacecraft design on a gravitational roller-coaster ride around Jupiter in 1974, finally rendezvousing with Saturn in 1979. NASA followed these probes with the *Voyager* missions, launched in 1977. *Voyager 1* visited Jupiter and Saturn, yet sacrificed its chance to visit the outer solar system by performing instead the first close flyby of Saturn's moon Titan. The enigmatic images of this huge, cloud-shrouded moon, with its pungent atmosphere rich in organic molecules, have remained a lasting legacy of this mission.

Voyager 2 hit the gravitational bull's-eye, sweeping past Jupiter in 1979, Saturn in 1981, Uranus in 1986, and Neptune in 1989. Spacecraft had only recently visited Jupiter and Saturn, and there were still many surprises in store. Upon racing by Io, the innermost of Jupiter's four largest moons, *Voyager 2* caught a plume of volcanic gas in silhouette. An astonished roomful of astronomers and planetary scientists immediately realized that Io was a geologically active world—the first to be recognized beyond Earth. Tremendous friction, generated as Io orbits its planetary parent, gives it heat.

Flying behind Saturn's moon Titan as viewed from the sun, *Voyager 2* caught its atmosphere in profile and revealed more of the complex organic chemistry on a moon that seems to have more in common with a low-temperature oil refinery than any atmosphere we have seen before. Beyond Saturn, it was all new: Uranus and Neptune revealed themselves as icy realms of gas and clouds. But flyby telemetry showed that each possessed a warm rocky interior and complex magnetic fields. Each planet unveiled to *Voyager* its previously poorly known retinue of icy moons, confirming the abundance of such moons among all the Jovian planets.

Voyager 1 and *Voyager 2* remain in operation today, almost forty years after their launch. They are around one hundred times as far from the sun as Earth is: a radio signal takes some seven hours to make the journey from their 3.7-meter radio antennae to the receivers of the Deep Space Network on Earth. Their science mission continues as they investigate the unknown boundary between the solar wind and the larger-scale environment of interstellar, as opposed to interplanetary, space. *Voyager 1* appears to have passed this boundary and is now in interstellar space; *Voyager 2* appears to still be in the process of crossing the poorly known frontier. Each will continue its journey out of the solar system. The warmth of their radioactive batteries will power onboard systems for at most another ten years. Neither is on course to pass close to any particular star. When their power dims, they will cease communication with Earth and go forth gently, and silently, into that good night.

Home Thoughts from Abroad

In traveling to the very edge of our home system, the twin *Voyager* probes have used their uniquely distant perspective to provide an unparalleled view of the solar system: a photographic portrait of our family of planets huddled around the faint sun. The images show both the fragility and the unity of the planets of the solar system: a distant cluster of figures sheltering around a small fire, surrounded by the darkest of nights. Irrespective of whether life exists beyond the pale blue dot of planet Earth, we can, after sixty years of space exploration, appreciate the diversity of the planets of our solar system even as we anticipate new secrets to be revealed by future missions. We have learned that several locations in the solar system, the planet Mars and the moons Europa, Enceladus, and Titan, perhaps offer something special: conditions under which life may arise. To them we now return and search in earnest.

Mars became a world on July 20, 1976—*Viking 1* arrived. The day began with a fiery streak high in the Martian sky as the lander— huddled within its protective heat shield—skimmed though the thin upper atmosphere. At six kilometers above the surface and traveling at nine hundred kilometers an hour, the spacecraft deployed a single sixteen-meter-diameter parachute. The kick from opening the parachute was not as fierce as you would think—the Martian atmosphere is only 1 percent as thick as Earth's—and it took some forty-five seconds for the parachute to slow the lander from nine hundred kilometers an hour to just over two hundred. At this point, the landing legs were extended and the multiple descent rockets fired, providing *Viking* with a controlled landing on the surface of Mars. Touchdown occurred at just before noon Earth time.[1]

The dust had barely settled before *Viking* captured its first image of the Martian surface and transmitted it to Earth twenty-five seconds later. The view that greeted Earth scientists eagerly watching the image appear, line by line, on the video screens in front of them was of a dusty rock-strewn plain. The gleaming white of the lander's hull and the sharp colors of the Stars and Stripes contrasted with the dull, blended reds of the Martian soil and rocks. Carl Sagan was a mission scientist with *Viking*, part of the group watching the first video images of Mars. His first impressions of the Martian landscape, later recalled in *Cosmos*, deserve to be quoted in full here: "I remember being transfixed by the first lander image to show the horizon of Mars. This was not an alien world, I thought. I knew places like it in Colorado and Arizona and Nevada. There were rocks and sand drifts and a distant eminence, as natural and unselfconscious as any landscape on Earth. Mars was a place. I would, of course, have been surprised to see a grizzled prospector emerge from behind a dune leading his mule, but at the same time the idea seemed appropriate."[2]

Human exploration of the Martian surface—using robotic landers and rovers—has continued apace in the four decades since *Viking 1* arrived. But no image has better captured the essence of a new world than Sagan's grizzled prospector surveying the landscape as seen by *Viking*. If we jump forward to 2012, we can meet Sagan's prospector face to face. The recently arrived *Curiosity* rover used MAHLI (the Mars Hand Lens Imager) on the end of its robotic arm to take an astonishing selfie. We see a grizzled *Curiosity*, already dirty and dusty from three months of prospecting on Mars, looking straight at us with its navigation cameras and laser system giving the impression of face and personality. The image is striking partly because of its resolution—the rover tracks, the windblown dusty soil, the faded distant hills are all picked out in crisp detail. The fresh sunlight of a Martian day reminds us of our own.

Humans have yet to set foot on Mars. But spacecraft from *Viking* to *Curiosity* have allowed us to experience vicariously the sensation of being there. From the science streamed back by a host of orbiters, landers, and rovers, we have discovered a world with a complex past and an enigmatic present. It is a fascinating story, the more so because the end has yet to be written: Mars lies at the very forefront of our search for life in the universe and shows both the excitement and the perils of astrobiology in action. So grab your mule and your shovel— it's time to go prospecting!

Waterworld

The idea that the Mars of today is but a dry relic of a once warmer, wetter Mars of the past has been fixed in our minds by the cumulative successes of a host of robotic Martian probes. So what evidence can we glimpse from orbit and scratch from the surface of Mars to convince ourselves of the ebb and flow of Mars's fluctuating fortunes?

It all began in 1971 with *Mariner 9*, the first spacecraft to orbit another world and send back images to Earth. The *Mariner* images resolved the surface of Mars into a network of what appeared to be channels and eroded gullies. These features were not Lowell's fleeting illusions. With each new generation of orbiting spacecraft— *Viking* (1976), the *Mars Global Surveyor* (1997), and the *Mars Reconnaissance Orbiter* (2006)—the surface of Mars has revealed an astonishing variety of deltas, alluvial fans, and outflow channels, each striking in its similarity to geological features on Earth driven by the action of water.

But no water itself. Not a drop. In addition, the surface of Mars is old: the density of impact craters, along with in situ radiometric dating of rocks performed by *Curiosity*, shows that the surface of Mars dates from over three billion years ago. Taken as a whole, the abundance of water-driven geological features viewed from Mars orbit points to the central role that water played in forming the ancient surface of Mars. But are these impressions supported by the ground truth—the geological results obtained by robotically roving geologists on the Martian surface?

NASA's fleet of rovers—*Sojourner* (1997), *Spirit* and *Opportunity* (2004), and *Curiosity* (2012)—has performed just that role for almost twenty years now. They are mobile geological platforms designed to investigate the physical appearance and detailed chemistry of rocks on the surface of Mars. Each carried a steadily improving array of magnifying lenses, grinders, drills (no hammers, unfortunately), and compact mobile laboratories to perform chemical analyses in situ.[3]

Rovers provide a close-up, almost microscopic view of the surface of Mars and its chemistry. Their only limitation is that they cannot dig—although *Curiosity* has a small drill—and they must seek out whatever interesting features are exposed to the surface. Like the orbiters, the rovers have discovered a surface driven by the action of water: fine wave-like ripples in sedimentary rocks, scatterings of granules of the mineral hematite (a form of iron ore), and clay beds— the last two of which, on Earth, are deposited by water-rich sediments.

At the same time, though, there is no smoking gun—no direct evidence of liquid water on Mars. All the orbital imaging and surface geology presents a strong but circumstantial picture of a wet ancient

Mars. Liquid water appears to provide the common link between the multitude of observations of the Martian surface. But we must always be on our guard against believing in the kind of stories we want to hear and the scientific explanations we would like to be true.

2001: A New Space Odyssey

So if water covered early Mars, where did it go? To my mind, the 2001 Mars Odyssey performed one of the most elegant observations of Mars. Odyssey is a Mars orbiter named in honor of Arthur C. Clarke's novel 2001: A Space Odyssey. It did not carry an intelligent supercomputer on a mission to investigate alien artifacts; instead, its primary mission was to map the surface of Mars by measuring the soft glow of energetic particles—neutrons—from the surface.

To understand why the surface of Mars should produce such a glow of neutrons, we need to take a further step back. Cosmic rays are energetic particles of extragalactic origin—they form a kind of low-level yet extremely high-energy background radiation field that permeates the universe. Collisions in the upper atmosphere destroy most incoming cosmic rays before they can reach Earth's surface. But if you lack an atmosphere, then these energetic particles bombard the top few meters of your planetary and lunar surfaces. Colliding with atoms in the surface rocks and regolith,[4] the cosmic rays produce a unique source of energetic, fast neutrons.

Nuclear reactors on Earth regularly produce such fast neutrons. In fact, controlling the flux of fast neutrons provides a way to regulate the power output of a nuclear reactor. The way to control fast neutrons in a reactor is quite literally to slow them down by using what is referred to as a moderator—a chunk of material made of atoms effective at slowing neutrons via repeated collisions—either graphite (carbon) control rods lowered into the reactor core or water (hydrogen) flowing around it. Odyssey observed this varying flux of moderated neutrons as they suffused from the surface of Mars. And remember—whatever was moderating the neutron flux had to be located in the top meter or two of the Martian surface.

The two most likely candidates for moderation on Mars are like-wise carbon and hydrogen: carbon in the form of carbon dioxide, and hydrogen in the form of water. Carbon dioxide is present on the surface of Mars, locked up as dry ice (solid CO_2) but restricted to the truly frigid polar caps. Over the remainder of the surface, hydrogen trapped in water provides the best (and not simply the most convenient) explanation of *Odyssey*'s observations. The conclusions are startling: lying under as little as ten or twenty centimeters of soil is a planetwide reservoir of water ice. On average, the material making up the first couple of meters of the Martian surface is about 14 percent water ice by mass. This is enough water to cover the surface of Mars in a layer of liquid water fourteen centimeters deep. Throw in the vast reserves at the pole and you are looking at a planetwide sea thirty meters deep.[5]

This all sounds really impressive—but seeing is believing. Inferring the presence of a huge subsurface ice reservoir from space-based observations is one thing; holding some of that liquid water in your hand (or robotic scoop) is another. The twin *Viking* landers found nothing—each was equipped with a simple scoop to re-trieve Martian soil to use in its experiments. But each could only scratch the surface, digging some five to ten centimeters into the soil. Furthermore, both *Viking* landers were located at Martian lati-tudes below fifty degrees—moderately close to the equator—in re-gions of the planet where the subsurface ice concentration was later observed to be low.

NASA's *Phoenix* lander saved the day in 2008. *Phoenix* was a static lander launched in direct response to the startling results from *Odyssey*. *Phoenix* was tasked with measuring the soil chemistry of the Martian surface, and not wishing to miss out on discovering the elusive subsurface ice a second time, NASA dispatched the lander to the frozen arctic plains of Mars, 68 degrees north of the equator. It was there that *Phoenix* imaged bright white ice in the soil while dig-ging a small sample trench. The ice disappeared a few days later, sub-limating into the atmosphere exactly as water ice would be expected to do.[6] Later, within *Phoenix*'s onboard lab, water vapor was detected boiling off the surface soil as it was heated within the spacecraft.

From that point on, it became easy to spot the ice—at least when you knew what you were looking for. Occasional meteorite impacts excavate bright, fresh ice, which then sublimates away over passing days, weeks, and months. The *Mars Reconnaissance Orbiter* captures images of these meteor impacts on average about once a year—such chance events showcase yet another example of time-lapse imaging of the Martian surface revealing unexpected new secrets.

Mars: The Dead Planet

Before I tackle the juicy question whether Mars once harbored life, or still does today, I want to impress on you the extent of our challenge. It does not matter whether we consider either the chemistry of the soil or the properties and composition of the Martian atmosphere—essentially, anything we can measure that is on or in contact with the surface tells us that Mars is dead, dead, dead.

The surface geology of Mars provides compelling evidence that it experienced a water-rich youth. Although the present-day water budget of Mars is believed to stretch to a global ocean thirty meters deep, orbital images of ancient surface features have led some researchers to speculate that Mars once had oceans up to five hundred meters deep.[7] Along with the evidence for ancient liquid water came the realization that the surface of the planet must have been warmed by an atmosphere considerably denser than that of present-day Mars—mainly because water requires specific conditions of temperature and pressure to remain as a liquid for any length of time. So what happened? Where did the atmosphere and all that water go?

We have come to realize that the loss of Martian water was bound up with the slow geological death of the planet itself. Mars is smaller than Earth, and one consequence of basic physics is that small planets lose their internal heat faster than large ones. Mars shows abundant evidence of massive volcanism: the ancient volcano Olympus

Mons—itself part of the vast, volcanic Tharsis plateau visible through even amateur telescopes on Earth—is the largest volcano in the solar system. Volcanoes allow gases trapped in the interior of a planet to vent to the surface, where they resupply the atmosphere. All evidence for erupting Martian volcanoes, lava flows, and geological activity, however, ceases some three billion years ago. Why? Because the interior of Mars cooled and solidified. The end of volcanism turned off the geological tap that replenished the atmosphere.

More critically, the cooling of the Martian interior weakened the planetary magnetic field—the one thing standing between the Martian atmosphere and the power of the solar wind. The unprotected atmosphere was literally boiled away into space by the blast of energetic particles in the solar wind. As the atmosphere diminished, so did its warming effect on the planet. The surface temperature dipped, and what water remained in the atmosphere and on the surface condensed and froze, creating the ice caps and subsurface ice layers discovered by *Odyssey*.

The News from Viking

More than at any other time in its exploration of the solar system, NASA threw the long ball with the *Viking* missions. Each carried a biological instrument package designed to test for the metabolic activity of Martian microbes and the presence of organic raw materials in the soil. If the twin *Viking* landers had discovered life on Mars in 1976, you can trust me that I would be telling you about it now.

Bottom line—the landers did not discover life. No signal was observed that departed from a sterilized control sample. At best there was a flicker of ambiguity when one experiment gave a result in line with life. It was the labeled release experiment in which a soil sample was doused with liquid water containing a mix of organic chemicals. The carbon in the nutrient package was dosed with radioactive carbon-14. The idea was that any living organisms in the soil would gratefully metabolize the organic nutrients and release radioactively labeled carbon dioxide as a by-product. This was exactly what happened; even further, a heat-sterilized control sample failed to produce any CO_2—exactly what would be expected if the heat had killed any living organisms. But when more nutrients were added one week later to the sample that produced the positive result, no more gas was produced. Were the microbes full after their unexpected meal (out from lunch)? Or had the original reaction been produced by a naturally occurring oxidizing material in the soil that, when used up in the original experiment, produced no further reaction? The result was ambiguous at best. Since all the other experiments flatlined and a number of possible nonbiological explanations for the positive result arose, the consensus was that *Viking* had not detected any evidence of life in the top few centimeters of soil at the two landing sites.

In many ways, though, this negative result shaped NASA's approach to Mars through the 1990s and up to the present day. *Viking* was a big, bold mission, and what followed was a more measured, incremental approach. The press replaced headlines of "looking for life" with the less brash "looking for past conditions that may have favored life." The small-scale yet progressive approach of faster, better, cheaper has been eminently sensible, equally exciting, and productive of a great deal of groundbreaking science. NASA has not given up on detecting life on Mars. But we have learned that you can't just land at a random location and expect to find it crawling into your scoop. We need to get smart and understand where limited local environments may defy the bleak and arid conditions that dominate the surface of Mars. We also need to remain optimistic—because I'm not done with the pessimism just yet.

The Very Modern Story of Methane, Lovelock, and Lowell

If you thought the surface of Mars was dead—well, the atmosphere is deader. What do I mean by "dead"? I mean that the mix of atmospheric ingredients on Mars is exactly what you would expect it to be, given the surface chemistry (which supplies raw materials—mainly CO_2 and water ice) and sunlight (which causes chemical reactions in the atmosphere). There are no chemical anomalies in the atmosphere that might point the finger of suspicion at life.

The idea that the atmosphere of a planet will show the chemical effects of living organisms has been most effectively popularized by James Lovelock, creator of the Gaia hypothesis, which views planet Earth as an integrated biosphere where the properties of the atmosphere are closely tied to life. The simplest example of this is the conversion of atmospheric and oceanic CO_2 into oxygen by planetwide photosynthetic organisms. In this sense, plentiful atmospheric oxygen is a biomarker that signals a living planet—a chemical whose abundance in the atmosphere is very difficult to explain through nonbiological reactions alone.

If all life on Earth were eradicated today, the oxygen in the atmosphere would remain for approximately two million years as it was slowly used up by oxidation reactions with surface rocks. Atmospheric methane—almost all of which is produced biologically—would be used up in just over twelve years: it is consumed voraciously by hydroxyl (OH^-) reactions in the atmosphere. In many ways, it is the presence of methane in Earth's atmosphere, at the level of a few parts per million, that is the more prominent biomarker—giving away of presence of present-day life on Earth.

So how would you react if I told you that methane had been detected on Mars in the small but chemically significant abundance of a few tens of parts per billion? It sounds even better when I tell you that the detection comes from three independent observations—from a spectrograph on the *Mars Express Orbiter* in 2004, from telescope observations on Earth in 2009, and finally from *Curiosity* in 2014. The observed signature of methane appears to vary rapidly over the space of a few months—now you see it, now you don't.

Some methane is expected in the Martian atmosphere, at the level of a few parts per billion. It is produced by the interaction of sunlight and trace amounts of organic chemicals delivered to the surface by meteorites. If confirmed, the existence of methane in excess of our expectations could point to a new phenomenon on Mars: either life exists there, which would be astonishing, or perhaps Mars's ancient volcanoes are not as dead as we think, which, given what we know of Martian geology, would be only slightly less astonishing (and would have its own implications for life on Mars). Perhaps the more troubling question is not so much what produces the methane, but where does it go? How does the atmosphere scrub itself clean in a matter of months? If the methane is real, I can guarantee you one thing—Mars is hiding some pretty big secrets.

The tale of methane on Mars may instead be a timely warning of being wary of the story you want to hear. The parallels with Lowell's observations of canals on Mars are worrying in their similarity. Lowell was essentially fooled into believing he had observed artificial canals on Mars by a combination of the limitations of his telescope and the blurring effects of Earth's atmosphere. In the same way, each of the claimed observations of Martian methane is worryingly flawed. The space-based claim requires that many individual observations be combined, and even then the claimed signature is barely discernible.[8] Observations from Earth are limited by the presence of abundant methane in our atmosphere (about one thousand times as much, atom for atom, as is claimed for Mars), through which we have to peer. It is cause for concern that observations from Earth claim to detect the presence of methane only when the motion of Mars with respect to Earth shifts the principal spectral signature of methane directly onto a strong methane line in Earth's atmosphere. In this case, the accuracy of any signal extracted for Mars is greatly reduced. When Mars is moving away from Earth, and the methane line is shifted into a clean part of the electromagnetic spectrum, no Martian methane line is observed.

Even the results from the *Curiosity* rover give cause for concern. From December 2013 to January 2014, the Mars rover measured a spike in atmospheric methane levels. Although an abundance of methane at ten parts per billion may not sound like much (and indeed isn't), the measurement was sufficiently in excess of previously measured background levels to be striking. The elevated methane levels were measured over a period of about two months before they declined sharply. Had *Curiosity* detected a seasonal bloom of methanogenic Martian microbes? Or was the methane outgassed by an unexpected volcanic vent?

Before losing ourselves in a symphony of speculation, we should perhaps listen to a few notes of caution. Unfortunately, the tunable laser spectrometer that *Curiosity* uses to test for methane in the Martian atmosphere is ever so slightly contaminated. Unwanted methane leaked in from Earth's atmosphere before launch, and small additional amounts are produced by the slow breakdown of onboard laboratory chemicals. How much? A few parts per million, which, when you are measuring a Martian signal of a few parts per billion, is a concern. None of this is news to *Curiosity*'s Sample Analysis for Mars team, and it has made strenuous efforts to remove such sources of contamination from the measurements. The low background level of methane, measured at a few parts per billion, seems secure. But the source of the short-lived methane spike, at a concentration ten times higher, remains the focus of intense scrutiny.

The jury is out. Or better put, the jury is busy searching for definitive evidence. *Curiosity* will continue to sniff the wind and subject any potential methane detections to more sensitive measurement techniques. In the meantime, two new satellites on the scene have much to contribute. Both NASA's *MAVEN* (*Mars Atmosphere and Volatile Evolution*) and the Indian Space Agency's *MOM* (*Mars Orbiter Mission*) probes arrived at Mars in late 2014—each with the goal of detecting the elusive signature of atmospheric methane. A further craft, the *Trace Gas Orbiter*, which will form part of the ESA's multimission ExoMars project, departed for Mars in March 2016. That is a great deal of investment—but getting a definitive answer to the question whether methane is present on Mars is both worth it and eminently achievable.

Should we be convinced by the current measurements of Martian methane? The existence of three independent claims is indeed compelling. But each claim must be judged on its own merits, and as we have seen, each gives cause for concern. Therefore, in the absence of a single, conclusive observation of what would be an incredibly important result for the possible biology of Mars, I for one remain skeptical. From our perspective as astrobiologists, it seems that—for the present at least—there is no convincing evidence for widespread Martian life in chemical contact with the atmosphere.

Okay—I have reached my personal nadir and plumbed the depths of pessimism regarding life on Mars. Now, how can we take all of the evidence and good, basic physics that I laid out above and break it? Does the bleak picture I have painted leave any loopholes in which life could exist on Mars?[9] One of my motivations in writing this book was to investigate the five most plausible scenarios for the discovery of alien life in the universe. You may have noticed that I am not doing a great job of selling Mars as a prospective habitat. So how can we cheat the system and fit life into a dead planet? How can we go out and find it? Let's try two approaches. First, can we answer the possibly easier question whether Earth life could survive on Mars? If so, what would this teach us about the prospects for indigenous Martian life? Second, can we get busy and look for nooks and crannies, environmental niches that buck the trend of Mars the dead planet and provide hidden or temporary havens for life?

Could Earth Life Survive on Mars?

Mars today is bleak, cold, and arid. Hostile to life, certainly—but lethal? Let's see. The surface conditions on Mars offer a multitude of ways to extinguish life: low temperature—dropping to a shuddering 180 Kelvin (–90° Celsius) at night; extremely low pressure—the surface pressure on Mars is equivalent to taking a stroll at approximately fifty kilometers above Earth's surface (putting your summit attempt of Everest at nine kilometers in the shade); a CO_2-dominated atmosphere; lots of solar UV radiation; cosmic rays, and apparently neither liquid water nor organic material in the soil at the few locations tested. Having said that, though, just how detrimental are these conditions? Can we distinguish between those that are life threatening and those that are merely inconvenient?

Surprisingly, temperature is not the critical factor that you might think it is—mainly because specific locations on Mars can be quite warm on any given day. On Earth, during the cold winter of 2014—dominated by the wonderfully named polar vortex—the media were keen to emphasize the harshness of the weather by pointing out which locations in North America were colder than Mars on any particular day. And yes, at a couple of locations in the United States and Canada, the temperature with the wind chill did fall below the average surface temperature of Mars (210 Kelvin, or −63° Celsius). But consider instead the real temperatures experienced by the Mars rovers on their wanderings over the Martian surface: maximum air temperatures peak at around 0° Celsius, and ground temperatures can reach as high as 20°.[10]

The reason for this apparent discrepancy is the difference between average as opposed to local temperatures. The average temperature on Earth is pretty much 15° Celsius—day in, day out. The nearer the equator you live, the more solar heating you receive, and the warmer you are; the farther from the equator, the colder you are. The temperature extremes on Earth (the difference between maximum and minimum) are smaller than on Mars mainly because of the thermal stability provided by the oceans and the lower atmosphere. On Mars there are big swings even on the best of days—when the sun sets, the temperature rapidly plunges to −90° Celsius. But for short periods each day near the equator, the sun can provide a potentially warm habitat for life.

The quickest way to determine whether any Earth organisms could survive and grow on Mars is to take a whole collection of them on an extended field trip to the planet itself. As we shall see, Mars possesses a wide range of local conditions that may be more or less welcoming of visitors—so to be thorough, you will have sow your seeds from Earth far and wide. But any thought of attempting this experiment brings with it a serious issue: as soon as we introduce Earth bugs to Mars, the potential exists for us to have contaminated the surface with new life, and thereby to invalidate all subsequent searches for truly Martian life.

This point is sufficiently important that space agencies such as NASA and the ESA spend a great deal of money ensuring such contamination does not happen, sterilizing all Martian spacecraft and storing them in "clean" rooms intended to be devoid of even the remotest spore or filament of life. Ironically, a species of resilient bacteria—in fact, the same species in each case—was found in both a NASA and an ESA clean room. Clearly, nature abhors a sterile room and is very determined to visit Mars. So how closely can you reproduce the surface conditions on Mars without leaving Earth? Well, there are two ways: book yourself a trip to Antarctica or make friends with a research scientist who operates a Mars environmental simulator.

Life in a Cold Climate

Tourist brochures recommend the dry valleys of Antarctica as the closest Mars-like place that you will find on Earth—and for once, reality lives up to the hype: These barren, rock-strewn valleys are true low-temperature deserts. Precipitation is low and falls as snow, which strong winds quickly scour away. The meager scratchings of soil are almost unbelievably poor. A scattering of ice-covered lakes is fed by the occasional melt of local ice fields.

Although the environment is cold, arid, and certainly hostile, it hosts—although not a profusion of life—a number of distinct, stable living communities. Perhaps of most interest to those searching for Martian life are colonies of photosynthetic cryptoendolithic bacteria, which live one to ten millimeters below the surface of primarily sandstone rocks.[11] They form communities within the tiny spaces available in the rock matrix itself, protected from the harshest winds, screened from the worst of the UV radiation, yet receiving sufficient sunlight to photosynthesize for a few months each year before freezing solid and hibernating through the long winter. The rocks heat up faster than the surrounding air and thus act as a kind of mineral greenhouse in which the bacteria can survive.

We still don't know how such organisms obtain their water—it could be that on the rare occasions when snow accumulates and melts, a bare minimum of water becomes available, which the bacteria harvest and store within thickened cell walls. Furthermore, organic nutrients are almost nonexistent, and it has been speculated that the bacteria obtain their phosphorus, sulfur, and the like from the rarest of wind-blown soil particles that get trapped in the rock.

Of course, even the relatively warm equatorial regions of Mars make the Antarctic dry valleys seem like a humid tropical paradise. But the dry valleys show that where there is even the most marginal physical support for life, there could be highly adapted, resilient organisms that can exploit such conditions.

Life in the Freezer

The second route to Mars is to re-create the surface conditions in a lab closer to home. You will need a pressure chamber that can be reduced to 1 percent of Earth's atmospheric pressure, into which gases are fed to match the composition of Mars. The interior of the chamber needs to be cooled with liquid nitrogen in order to simulate the chill of the Martian surface, and sunlight is mimicked by a powerful, full-spectrum arc lamp. A number of research labs around the world maintain Mars environment simulators, and in this way Earth organisms—mostly microbes—can be introduced into a simulated Martian environment and their responses measured by a variety of tests for growth, metabolic, and genetic activity.

What have we discovered? It turns out that a number of species of Earth microbes can survive and grow in conditions of Martian temperature, pressure, atmospheric composition, and solar irradiance. It is interesting that not all such winners in the game of Martian survivor are classic extremophiles. A number of species of the bacterial genus *Carnobacterium* do rather well under simulated Mars conditions. Such bacteria can often be found growing within containers of refrigerated vacuum-packed meat—hence their name—and it was later discovered that their natural habitats include the oxygen-starved depths of the Siberian permafrost. A particular group of species actually fares better under simulated Martian conditions than at normal Earth temperature and pressure—clearly, these guys are first in line for the trip to Mars!

The inescapable fact, though, is that Earth life must have access to liquid water—even tiny amounts—in order to survive. The rock-living bacterial colonies of the Antarctic dry valleys show just how preciously little food and water Earth life requires to exist. It is not much, but it still needs something. Because of that fundamental requirement, I can't give you a definite answer whether Earth life could survive on Mars. But if we can locate even the most isolated or transitory pockets of liquid water—the most marginal amount of melt of the abundant subsurface ice—then there exists the possibility that Earth life could exploit it. And if Earth life could—then why not Mars life? But what chance is there of discovering what may be a rare, highly localized, and seasonal occurrence on (or below) the surface of Mars from either our orbital lookout posts or within the limited range of a surface rover?

A New Hope

Mars is riven with gullies. Each new generation of orbiting observatories around Mars has discovered gullies, rilles, and channels of yet finer detail and scale. But no one expected to see what appeared to be seasonal flows within these gullies—growing in the Martian spring and receding in the fall. And if that did not come as a sufficient surprise, the real eye-opener was to see these flows recur in the same locations year after year. It turns out that the surface of Mars is restless, active, and dynamic—we simply required orbital cameras of sufficient resolution and scientists with the discipline and patience to discover it. Once again, the *MRO*'s HiRISE camera provided the vision with which to resolve a new level of detail on the Martian surface.

The features themselves are referred to as recurring slope lineae (abbreviated as RSL in the academic journals): dark yet discernible bands running down small gullies (as small as one to twenty meters wide). The science team at the University of Arizona's Lunar and Planetary Lab, which developed the HiRISE camera, first presented the RSL features in a 2011 research paper. It then followed up with a more comprehensive study published in late 2013. These dark stains grow incrementally, advancing slowly with the warmer seasons—northern sites follow the northern spring and summer, southern sites follow the southern seasons—and can be easily distinguished from the more abrupt dust and rock avalanches that also occur (though typically at different locations).

The RSL often appear below distinct layers of bedrock exposed along the rims of larger valleys and craters. Critically, they occur on moderately steep slopes, with inclines greater than thirty degrees. Such terrain presents a considerable challenge to our current generation of rovers (remember that the redoubtable Mars rover *Spirit* succumbed to a small sand dune). This is unfortunate because these slopes forming recurrent seasonal gullies are our top location for seeking out present-day life on Mars.

Why? Water—most likely, salty, liquid water—presents the best explanation for the formation of these features. Just as when you water your garden in summer and the bare soil darkens when wet, the growth of dark bands in Martian gullies is taken as evidence of the flow of liquid water leaving a moist trail in the soil. Most seasonal-gully features occur in the slightly darker rock terrain of the Southern Hemisphere and in the equatorial Valles Marineris, where measured ground temperatures are high—though in this case, "high" means a temperature greater than 250 Kelvin, or −23° Celsius—and therefore where salty water, rich in brines, can exist as a liquid. Moreover, their recurrent nature provides a potentially stable habitat; year on year, whatever meager source of sustenance might be available to Martian organisms can be concentrated and replenished there.[12]

I warned you earlier that Martian methane might well be the story that we want to hear rather than the one that the data support. Am I falling into the same trap here? Well, the RSL features are real and, as far as we can tell, different from the other variable features we have discovered on the surface of Mars. Is it definitely water? No, clearly not. The *MRO* carries a spectrometer capable of detecting the spectroscopic signature of water if present. There currently has been no such detection associated with these dark bands (although the signature of salt deposits in these gullies has been detected). And there remains considerable uncertainty about how much water may be present in these gullies—only a tiny fraction of salty water in the soil may be enough to produce the dark features.

There also remain a couple of sticky questions concerning the water supply at the current RSL sites. They appear to be present mainly at low latitudes, where there is a greater amount of solar heating. But *Odyssey* told us that the abundance of water ice in the Martian surface is lowest at such latitudes. Once again, is only a tiny amount of melt required to produce the observed features? More troubling, though, is the fact that these features occur over successive years. If we are indeed witnessing the seasonal melting of small amounts of water ice, won't the local ice reservoir soon run dry? Given the great age of the Martian surface, the fact that there are any RSL sites visible today tells us that if they are indeed linked to liquid water, they are being replenished in some way. How exactly this might happen is pure speculation: it could be from atmospheric water vapor captured via reactions with the Martian soil. More intriguingly, these local melting sites could be the mere tip of the iceberg[13]— part of a more complex underground aquifer system where liquid water flows underground, guided by strata of impermeable rock. Is the fact that the RSL appear to be associated with exposed layers of bedrock significant in this respect? We just don't know.

Finally we must consider how the potential habitat provided by these seasonal gullies fits in with our broader picture of the evidence for surface life on Mars. *Viking* said no to present-day life on Mars— or at least a firm "I don't know." But the *Viking* landers visited the rolling plains, suitable for a safe, automated landing, rather than the more challenging gully terrain (craters and valleys) containing the RSL sites. We have already learned from the Antarctic dry valleys that in challenging environments, the distribution of living organisms is linked closely to availability of even the smallest amount of liquid water.

We have also come to realize that the chemical composition of the Martian atmosphere suggests that there can be no widespread surface (or slightly subsurface) life on Mars that makes biological use of it (either feeding from it or contributing to it). The word "widespread" might contain the key point: the potential number of steep-slope sites with the right local geology and slope conditions amounts to less than 1 percent of the Martian surface. Local environmental conditions may place strict constraints on the abundance of Martian life—restricting it to a level that keeps it below the atmospheric radar.

What is clear is that the discovery of seasonal gullies on Mars has raised perhaps many more questions than it has answered. I am sure that the team behind the *MRO* and HiRISE is probably overjoyed with this situation—discovering new, currently unexplained facts about nature is exactly how scientific progress occurs.

One thing I can say is that if we are indeed witnessing the seasonal melting of a salty ice layer and the consequent formation of recurrent moist bands in the soil of these Martian gullies, then we have discovered exactly the kind of potential habitat for life that we have been searching for. We have studied examples of Earth life that can survive and grow in Martian surface conditions. At this point, liquid water remains the key ingredient that we are lacking. With the discovery of these RSL sites, we have identified a number of locations on the surface of Mars where there is plausible evidence for the presence of liquid water.

I started this book by giving you the current bottom line—there is no scientific evidence of life beyond Earth. But in the seasonal Martian gullies we have discovered the first of our five plausible scenarios for where that life could be found. But I would like to do more than just hint at life. How will we get there, what form might such life take, and what science will be required to confirm its presence?

ALH84001

A good general life lesson is to be careful what you wish for. Astrobiologists dearly wish for a sample return mission to deliver a sizable amount of Mars to their Earth laboratories for detailed study. In 1984, this wish was granted—with long-lasting consequences. In that year a dark rocky meteorite was recovered from the Allan Hills region of Antarctica—receiving the identification code ALH84001. For fairly obvious reasons, Antarctica offers a number of advantages to the budding meteorite hunter: most of the continent is white, and most meteorites are black.[14]

ALH84001 is an SNC meteorite, a member of the geological class of meteorites that have been determined to be of Martian origin (from the precious subset of these meteorites that preserve trace amounts of the Martian atmosphere in their mineral matrix). It weighs just under two kilograms; if sold at current market prices of $1,000 per gram, it would net you some $2 million to fund your Mars exploration program.

In many ways, though, ALH84001 is priceless. Unique among Martian meteorites, it defines a group unto itself within the SNC classification. ALH84001, a rocky meteorite, is the oldest-known chunk of Mars in our possession, with a radiometric age of just over four billion years.[15] This means that the rock of which ALH84001 is composed was laid down during the earliest periods in Mars's history, when the surface geology was dominated by liquid water—the "warm, wet phase." Indeed, within the mineral matrix are deposits of carbonate minerals that—at least on Earth—condense out of warm liquid water. Unfortunately for ALH84001, yet to our considerable benefit, these rocks did not remain on Mars: a separate meteorite impact blasted it from the surface of Mars some fifteen million years ago and left it floating through interplanetary space. Our paths crossed thirteen thousand years ago when ALH84001 made its fiery descent to the ice fields of Antarctica, where it began its slow journey across the continent.

Twelve years after its discovery, ALH84001 became big news. The NASA-led team tasked with studying the meteorite completed a long and careful study. The resulting academic paper, titled "Search for Past Life on Mars: Possible Relic Biogenic Activity in Martian Meteorite ALH84001," was accepted by the journal *Science*. If that did not get your astrobiological feelers twitching, then NASA made sure you were aware of the potentially stunning consequences of this discovery by organizing a major press conference.[16]

Let's focus on the science rather than the furor that followed. What hidden evidence did the team uncover in ALH84001? Each piece of evidence is associated with the carbonate minerals present in the cracks and fissures of the volcanic rock. The first clue is that the carbonates contain a group of organic compounds called polycyclic aromatic hydrocarbons (PAHs). The team concluded that these PAH compounds were preferentially associated with the interior of the meteorite and were not simply a contaminant picked up during its many thousands of years on Earth.[17] In terrestrial fossils, these can be (but are not always) associated with decayed biological matter. Second, the carbonates contain distinct deposits of specific iron- and sulfur-containing minerals, and this mineral zoning can occur in some terrestrial organisms that metabolize such metal compounds. The third piece of evidence is really neat, and potentially the strongest thread in this story: clustered near the edges of the carbonate minerals are tiny ordered crystals of the mineral magnetite (Fe_3O_4). In their size, shape, and exquisite detail, these crystals are very similar to those produced by magnetite-processing bacteria on Earth. Finally, the most photogenic of all the pieces of evidence presented for ALH84001 is the claimed presence of bacteria-shaped objects (unimaginatively called BSOs for short) on the surfaces of internal fractures within the meteorite. These tiny nanofossils look very similar to terrestrial microfossils—except that for the most part they are much, much smaller. Taken together, the team that studied ALH84001 argued that a biological origin—ancient Martian microbes—for these four lines of evidence provides the best single explanation of the features found within the meteorite.

Now the controversy could unfold. Within the field of astrobiology, one of Carl Sagan's most repeated maxims is that "extraordinary claims require extraordinary evidence." Put another way, when you claim a major discovery, your scientific evidence and your conclusions based on it must be able to withstand extraordinary levels of scrutiny and come away unscathed. It did not take long for that scrutiny to begin.

One external scientist brought in by NASA's director, Dan Goldin, to provide an independent assessment of the agency's claims was Bill Schopf. Schopf was an astute choice: an acclaimed paleobiologist and the author of numerous papers investigating the evidence for ancient fossilized life on Earth in Archean and Proterozoic rocks.[18] Schopf was not allowed to perform his own analysis, merely to assess the published evidence plus the team's conclusions and then to provide his own perspective. His initial assessment—somewhat sidelined in the rush to produce high-impact newspaper headlines—has stood the test of time: each of the four pieces of evidence presented to support the case for ancient fossils in ALH84001 could be the result of plausible nonbiological effects or of terrestrial contamination from its thirteen-thousand-year layover in Antarctica. Martian biology remains a plausible explanation of the features observed in ALH84001. But it is only one of a number of explanations, and the quality of the data does not provide a clear indication of who is right.

Having said that, though, I think there is an important broader lesson to be learned from the story of ALH84001: the scientific analysis applied to this meteorite has been excellent—both in the original paper and in the longer-term follow-up work conducted by independent teams. In hindsight, the research teams would probably not do anything differently (though perhaps they might rein in the media machine). But I think this story provides a showcase of *exactly* how we would treat rocks and soil obtained from a future sample return mission. In this sense, ALH84001 and the scientific analysis applied to it remain of exceptional value. It offers a warning that a future sample return mission to Mars may not offer the clear answers that we perhaps naively expect. We won't know for sure until we try. All we can wish for now is a little more Mars material—preferably obtained under scientifically controlled conditions directly from the Martian surface.

Aim High, Throw Long

It has been nearly twenty years since Mars exploration resumed in earnest following the long hiatus after the *Viking* missions. We have learned a lot in that time: ancient water-driven geology, present-day subsurface water ice, soil chemistry, and seasonal surface changes. More important, we have learned to frame new questions: Do ancient sediments preserve Archean-type fossils? Is liquid water associated with seasonal gully formations? Could subsurface ice provide a habitat for life? Perhaps most critical of all, we have learned how we should answer these questions: via a bold sample return mission to Mars.

A successful sample return mission will permit exactly the kind of leap in our understanding of life on Mars—both ancient and modern—that astrobiology requires. This is not just my view: in 2013, the U.S. National Research Council produced its decadal report detailing the national aims for solar system exploration over the next ten years. Its top priority was to prepare for a sample return mission to Mars—correctly identifying it as the most effective way to obtain comprehensive answers regarding the surface conditions on Mars and its potential as a habitat. The report was requested by both NASA and the National Science Foundation (the source of much U.S. government research money) and so carries a lot of weight. But as we shall see, good recommendations are one thing, good decisions based on them (taken in the harsh light of limited budgets) quite another.

Our Best Shot at Life on Mars?

We have talked a lot about Mars and its potential to host life. It is now time to get specific about where that life may reside and how we will perform real tests to confirm it. Goal number one is to look for ancient fossils within sedimentary rocks. While this may not be as exciting as looking for present-day life, the science looks rock solid and the practical challenges appear surmountable.

The *Curiosity* rover is currently exploring exactly the kind of terrain we are interested in—exposed ancient rocks (contemporary with ALH84001) whose geological environment points to its having been immersed in warm liquid water in the distant past. I should also point out that *Curiosity* is not discovering these regions by chance—its landing site was selected from a list of candidate locations studied in exceptional detail by the *MRO* to identify exactly the kind of conditions the rover is now investigating. *Curiosity* is able to drill small core samples from the rock itself—to a depth of about five centimeters. Although scientists might ultimately want to drill a little deeper into the surface—and return solid core samples rather than powdered rock—all *Curiosity* really lacks at present is the means to return the samples to Earth.

The time period for the water-rich early history of Mars coincides exactly with the period when life was just getting going on Earth, between about 3.5 and 4 billion years ago, just after the end of the late heavy bombardment. The search for ancient fossilized life on Earth employs exactly the kind of methodology we should use to test this idea: microscopic searches for cellular order coupled with biochemical tests for metabolic products and the residue of decayed organic matter. To perform such a search we will have to obtain rock specimens sampled from a range of terrains likely to contain such fossils—ancient sedimentary rock whose formation chemistry points to a benign, stable, liquid-water environment—and then return them to Earth.

Given this challenge, we can see the succession of rover missions sent to explore the surface of Mars in their astrobiological context: slowly but surely we are revealing the kinds of sites where we believe the conditions approximately 3.5 billion years ago may have been exactly right for life to form. In fact, given that such a large amount of the Martian surface is composed of such ancient rocks, one possibility is that we may end up learning more about the origin of life on Mars than we currently know about the origin of life on Earth.

A second goal will be to return samples from the enigmatic seasonal gully formations. Here we will be looking for evidence of an active ecosystem as opposed to a fossilized relic. It might be (extremely) optimistic to expect living organisms to survive the return journey to Earth. Instead, returned samples would be analyzed for evidence of microscopic biological order and for clues about whether the soil chemistry either provides a source of metabolic energy or contains possible by-products. However, the seasonal melt gullies (if that is indeed what they are) present some of the unfriendliest ground to your aspiring rover: steep, loose terrain at the head of rock walls. Timing is also an issue: how long do seasonal formations persist? Do gullies dry up periodically or cease altogether?

Clearly, although the prospect is enticing, the challenge of returning samples from Martian gullies remains high, even given NASA's track record for innovation. You could go for the more practical option of drilling down a couple of feet to the water-ice layer and seeing what you find. But be prepared for disappointment if you just dig at random—there may be no melting, no concentrated organic material, and no life. But the ice layer is accessible and abundant, so you would clearly do well to include samples of it in your return package.

Beeps, Scoops, and Leaves

How would a sample return mission to Mars proceed? In many ways it would be similar to a relay race: each technical challenge associated with the mission—orbiter, rover, ascent vehicle, and Earth-return vehicle—would be undertaken by an independent craft with the precious Martian samples carefully passed through each stage in a well-choreographed baton exchange. We are familiar with some components of the mission—the dispatch of an orbiter and a lander to work in tandem in exploring the surface for the best sites. A rover similar in design to *Curiosity* might well identify and cache a number of samples, collected from a range of interesting terrains. The mission could take a number of years, involving long journeys across the surface, and the rover would require a wide range of sensors and equipment to identify and collect samples of the greatest interest. In short, a big, beefy rover.

How would the samples return to Earth? This would likely be a two-stage process. First, a Mars ascent vehicle would carry the sample from the surface of Mars to a low-Mars orbit. In fact, transporting the Mars samples from the surface to orbit is the key link in the whole mission chain—gravity is not on your side, and you have to pay for every kilogram carried aloft. Twenty kilograms is probably on the low end of the mass of material you might want to return (remember, though, that this payload weight includes the highly durable sample containers as well). To lift this material into orbit, you have to accelerate it to a velocity of almost 5 kilometers per second—the escape velocity of Mars.[19] For a modest twenty-kilogram payload, you would not need a huge rocket: current designs feature a four-hundred-kilogram rocket, about four meters long and with a diameter of about half a meter. The real challenge is to make sure that that rocket functions after six months in deep space, the rocky ride to the surface, and up to a year parked on the surface of Mars—quite clearly, you only have one shot, so you have to make sure it counts.

You might harbor dreams of a larger payload mass, say, a bulky 200 kilograms of Martian material. But the rocket required to bring back such an amount starts to become an unwieldy beast—up to 1,500 kilograms and six meters long. That is a lot of rocket—plus a modest ground station. The mass of *Curiosity* is 900 kilograms, so you are looking at a considerable challenge in getting a large ascent rocket safely onto the surface of Mars. So perhaps 20 kilograms of Mars is the best you can hope for. That total sample could consist of a number of packages selected from a range of surface terrain—still scientifically very exciting.

Once in orbit, the ascent vehicle must perform an automated docking maneuver with the mission orbiter and hand over the precious cargo. The orbiter or a conjoined craft would then act as the Earth-return vehicle—using a rocket motor to return to Earth, enter orbit, and return the sample container to the surface. This is a delicate phase: both NASA's *Stardust* and the Japanese Space Agency's *Hayabusa* probes returned samples of cometary and asteroid material to Earth along very-high-velocity trajectories—12 kilometers per second in each case. That is both a lot of energy to burn off as you reenter the atmosphere and a potentially risky way to return your Mars samples to Earth. If the parachute fails and the return capsule ruptures, then Houston, we all have a problem!

Affordable options exist for a lower-risk return to Earth. The commercially operated—and reusable—low-Earth-orbit *Dragon* capsule could be used following a standard cargo mission to ferry the sample container to Earth. Those of you who have read *The Andromeda Strain* are aware of the raft of issues associated with returning extraterrestrial material to Earth—especially samples that may contain biological material. Needless to say, there are also good scientific reasons why you would not want the Mars material to come into contact with anything from Earth—you will have gone to a lot of trouble to keep your samples of Mars pristine, and you won't stop when you get them back to Earth. So the samples will end their journey in a specially prepared analysis facility, and then the science will begin.

Harsh Reality

This all sounds great—but will it happen? Back in 2005, NASA and the ESA realized that the challenge of a sample return mission to Mars would stretch beyond the capabilities of a single space agency. Think of it this way: each component of the proposed mission—orbiter, rover, and Mars ascent vehicle—represents a flagship-class mission in its own right, that is, one with a budget significantly greater than $1 billion. When you add in the potential need for a dedicated sample-handling facility on Earth, you are looking at a total budget that may exceed $7 billion. But what began as a promising partnership to cooperate on the multiple coordinated missions required for sample return was scuppered by a 20 percent cut to NASA's robotic exploration budget in 2011, with little prospect for future improvements.

Harsh financial reality forced NASA to shelve its commitment to sample return. None of the science has gone away—nor the scientists who believe in the knowledge we will gain from a sample return mission. NASA and the ESA are each now planning to send its own *Curiosity*-class rover to Mars in the years leading up to 2020. With little money to spend on solar system exploration as it is, this duplication of effort seems to be a pretty inefficient use of limited resources.[20]

Since my $4 billion will not stretch to cover a sample return mission, is this how I should choose to apply my science funds? Would more rover missions be my top route to discovering alien life in the solar system? To be honest—no (to be fair, searching for life is not NASA's sole priority for robotic exploration). If we were talking about a combined effort to really nail a sample return mission to Mars once and for all, then I would be willing to take the risk. Any lesser goal for Mars would just not make it as my flagship astrobiology mission—especially given the surprises that the farther reaches of the solar system have in store for us.

I originally gave you $4 billion to fund your astrobiology research—albeit as a playful figure rather than a well-worked-out budget—and now I think you can see the balance that scientists and science funders must strike. Would you seek out another reader of this book (the European edition), pool your funds (all of them), and place effectively all of your eggs in a Mars sample return basket? Fortunately, you have a few more chapters to read while pondering your answer—mainly because it seems unfair to push you on this question before we have really delved into all the options in front of us.

Human Exploration of Mars

Before we bid farewell to Mars, I want to offer a parting thought: are human missions to Mars just a gimmick, one small step and a flag-planting exercise, or do they offer one giant leap for clearing the obstacles that slow our pace of robotic planetary exploration?

Consider two views. One holds that there is no scientific task that a robot cannot perform as well as or better than a human—while requiring less food and positive feedback in the process. Quite literally, robots carry less baggage—emotional and physical. They consume nothing but electrical power and produce no biological waste. HAL 9000 aside, robots don't get depressed, lose bone mass in low-gravity environments, or suffer the debilitating effects of radiation sickness due to cosmic rays and the solar wind.[21] While a sample return mission to Mars may require approximately two tons of equipment to be landed on the surface and a maximum of twenty kilograms to be returned to Earth, a human mission to Mars may require a minimum of forty tons of material to be landed on the surface. Don't even get me started on the cost!

But there is an alternative view. In 1972, Harrison Schmitt became the first scientist to visit the moon when he journeyed as an astronaut on Apollo 17. He and fellow astronaut Gene Cernan visited the Taurus-Littrow region of the moon—part of the ancient lunar highlands with evidence of volcanic activity. It was on his last moonwalk that Schmitt, a geologist by training, spotted an "odd" rock, a piece of volcanic ejecta that did not fit with its local surroundings. The rock, more formally known as Troctolite 76535, has been described as the most interesting sample ever returned from the moon for the wealth of geological information it contains. His discovery raises a big question for robotic exploration of the solar system: how do you train a robot to replicate Schmitt's field experience and talent for noticing the odd rock in a thousand, or at least provide a remote operator with the same sensory experience as an astronaut on Mars?

In some ways, these are purely hypothetical questions. Should humans visit Mars, they will do so for more than scientific reasons— the same way that we came to visit the moon. All I know for sure is that my $4 billion won't cover it!

In one hundred years, people may well question why astrobiologists began their search for life in the solar system with dusty, dead Mars and not with the salty oceans of liquid water encased within the ice moons of Jupiter and Saturn. Hang on a minute, I hear you say! Am I telling you that we have been scratching around in the Martian regolith, looking for an occasional bead of moisture when there are vast lunar oceans in the outer solar system? Well, yes—and you have every right to feel aggrieved that it has taken us so long to get here. Though I should note that you are not alone, since there are numerous planetary scientists who have argued for the habitability of the Jovian moons for many a long year.

The Galilean Satellites

One of the most engaging astronomical experiences for any observer of the night sky—great or small—is to look toward the giant planet Jupiter as it makes its stately way around the outer solar system. If the sky is dark and your timing fortunate, even a small telescope will reveal four faint companion stars, strung out in a line, accompanying Jupiter. These are the stars first noticed by Galileo in January 1610. His telescope was not powerful enough to reveal more than their positions. But during two months of careful observations, he noticed their cyclical movements about Jupiter. Galileo's conclusions, published in April 1610 as part of his *Starry Messenger*, were astounding: four moons accompanied Jupiter. They had the fateful temerity to defy astronomical (and theological) doctrine and pay orbital homage to Jupiter rather than to Earth. Although his original name for the newly discovered moons—"Cosimo's stars," after his Medicean benefactor—proved less durable than "Io," "Europa," "Ganymede," and "Callisto" (as suggested by Kepler), these four new worlds were destined to play a prominent role in both future science and imagination.

When did the Galilean satellites become worlds? The 1970s brought a flurry of brief encounters as *Pioneer 10* and *Pioneer 11*, followed by *Voyager 1* and *Voyager 2*, made high-velocity flybys of Jupiter and its inner moons. Rather like the trajectories, the views of the Galilean satellites were fleeting, tantalizing affairs. The *Pioneer* probes performed relatively distant flybys of the moons and returned only low-resolution images—enigmatic views that offered only a marginal improvement over images taken by telescopes on Earth. The *Voyager* probes performed better, achieving close flybys of Io and Europa.[1] *Voyager* showed Io as a young and restless volcanic world whose rocky surface was paved with sulfur-rich lava fields. For the first time, we saw plumes of ejecta caught in silhouette against the backdrop of space—evidence that active volcanoes were present beyond Earth. In comparison with the fiery cauldron of Io, Europa could not have appeared more different, its surface smooth and unblemished—a hard sheen of water ice encasing an unseen rocky interior.

Beyond the orbit of Europa lie Ganymede and Callisto—the giants of the Jovian lunar system. Ganymede is the largest moon in the solar system, a shade larger than the planet Mercury—though, interestingly, it possesses less than 50 percent of the mass of Mercury because of its mixed composition of ice and rock. Callisto is often unfairly labeled the "boring moon" of the Galilean satellites. Slightly smaller and less massive than Ganymede. Slightly more cratered, with slightly less evidence of surface reprocessing. Callisto fits our idea of the Jovian satellite system: as you get farther from Jupiter, each moon appears to be a less dynamic, less vigorous lunar environment.

Each of the ice moons—Europa, Ganymede, and Callisto—is dusted with a patina of what can best be described as pinky-orangey-brown stuff. Exactly what this stuff is made of is unclear. It could be salts and minerals brought to the surface by ice welling up through fissures. It could be a smattering of organic material deposited on the surface along with interplanetary dust. It could be both—or neither. Astronomers have observed it very carefully, but the material presents no clear spectral signature—which is completely consistent with both of the above explanations. So until we go there and scrape some off, we are unlikely to get definitive answers to this interesting conundrum.

Sleight of Hand

If you were paying attention earlier, you noticed that I skipped rather too quickly over a suspicious observation. Why is Europa so smooth? Even today the solar system is cluttered with tiny debris that we call comets and asteroids—the leftovers from the long-gone days of planetary birth. They continually pepper the surfaces of the airless worlds of the solar system, and their cratered remains accumulate across each planet and moon with the passing of time. The only way to erase these craters is to create a new surface—in Io's case by active volcanism and the spread of lava fields. But why should Europa possess a young, uncratered surface—as young as fifty million years old (which is the mere blink of an eye for a long-term solar system resident)? Even with the fleeting views of Europa provided by the *Pioneer* and *Voyager* flybys, astronomers and planetary scientists observed that although uncratered, the surface of Europa was in places cracked and chaotic, as if disturbed by deep geological forces.

But "geology" isn't exactly the right word to use here. "Cryogeology" is a better term—describing a system of forces on an icy analogue of Earth in which water plays the same geological role as rock. This view of Europa as a cryo-geological world substitutes water ice for a hard rocky surface; the surface ice floats on a "magma" of either warmer ice or liquid water, and is resurfaced by upwellings and plate tectonics. It was a startling and ultimately correct view of Europa. The speculation that Europa might harbor liquid water was potentially extraordinary in its implications for life—and one that was confirmed by a later visitor to the Jovian moons.

A New Galileo

In 1989, a bold new mission was launched to the outer solar system. *Galileo* would be the first probe to enter the orbit of Jupiter, following a long and winding six-year route through the inner solar system as it gained gravitational energy before finally passing through the asteroid belt en route to its target. The elegant maneuvers executed by *Galileo* are referred to as VEEGA (Venus-Earth-Earth gravity assist). Rather like the slingshot trajectories taken by the *Voyager* probes, each planetary encounter was tuned to boost the spacecraft's velocity in exchange for an infinitesimally small amount of each planet's orbital energy.

The *Galileo* spacecraft probably experienced the most scientifically interesting premission phase of any probe in our short history of planetary exploration: at the behest of Carl Sagan, *Galileo* powered up its onboard instruments during the slingshot flyby of Earth in 1990 and performed what could be called the first astrobiological survey for life on Earth (an idea to which we return in chapter 8).

Galileo's modern voyage of discovery continued through the asteroid belt en route to the outer solar system, where it discovered the first moon accompanying an asteroid—the tiny rock Dactyl was observed in orbit around the slightly larger rock Ida. This discovery, combined with its bird's-eye view of comet Shoemaker-Levy colliding with Jupiter in 1994, meant that *Galileo* could be said to have had a pretty successful journey even before arriving in Jupiter orbit in December 1995.

Down, Down, Down to Business

Galileo carried with it a descent probe designed to separate from the main spacecraft and make a supersonic plunge into the unknown depths of the Jovian atmosphere. The probe separated from *Galileo* in July 1995, and on December 7 began its descent phase. Atmospheric entry was staggeringly fast—47 kilometers per second (roughly, 169,000 kilometers per hour)—and the frictional deceleration experienced by the probe was utterly uncompromising: subjected to 230 g of deceleration,[2] the probe slowed from its impact speed to subsonic in only two minutes. Just over half of the 150-kilogram heat shield was vaporized in the fiery descent.

At this point, the descent mission nearly ended in disaster. The craft was supposed to deploy a parachute, which would slow the probe to a more stately 160 kilometers per hour and allow for more detailed atmospheric measurements to be recorded. But the parachute did not open on time, deploying one agonizing minute later than planned. By rights the parachute should have failed to function completely: an accelerometer controlling the parachute deployment had been installed backward, and exactly what did allow the parachute to deploy remains a mystery. On the parachute's opening, the probe descended 156 kilometers into Jupiter's atmosphere, rewarding the mission scientists with just under one hour's worth of data, which revealed a turbulent, chemically rich, and dynamic world.

The probe eventually succumbed to the rising temperature and pressure of the Jovian interior: first the parachute melted, initiating a long free fall into the depths of the planet. With no solid surface to strike, the components of the probe steadily melted and then vaporized one after the other until the individual atoms ended their mission by mixing with the liquid metallic hydrogen of Jupiter's core.

Jupiter Ho!

Galileo entered Jupiter orbit on December 8, just one day after the drama and revelations of the descent probe. Although *Galileo* was a Jupiter orbiter, the proximity of Io, Europa, Ganymede, and Callisto offered the opportunity for multiple flybys of the Galilean satellites. During its eight-year mission to Jupiter, *Galileo* completed thirty-five orbits of the planet and encountered Europa eleven times (which in hindsight seems a small number given the amount we learned). Perhaps the most important of *Galileo*'s suite of instruments was its magnetometer—two sets of three detectors strung out on an eleven-meter boom to isolate it from the latent magnetic field of the craft itself.[3] It was designed to probe Jupiter's vast magnetic field—second only to the sun's—which is whipped up within its vast conducting core of liquid metallic hydrogen.

What no one expected was that Jupiter's magnetic field would in turn create (the more correct term is "induce") magnetic fields within the Galilean satellites themselves. Most surprising of all was Europa. Though Europa is relatively large for a moon—slightly larger than Earth's moon—any heat left over from its formation should have long since cooled, leaving a solid, rocky core. What *Galileo* discovered was both unexpected and elegant in equal measure—Europa possesses a weak magnetic field. The fact that it is induced by Jupiter and is not intrinsic to Europa itself is obvious when you consider the following three facts: Europa's magnetic field rotates once every eleven hours; Europa itself rotates once every 3.55 days; Jupiter rotates once every eleven hours. Jupiter was clearly the driving power behind Europa's magnetic field. To induce a magnetic field in a moon such as Europa requires the presence of a lunar-wide conductor. In the case of planets, the ability to conduct electrons is provided by some form of liquid-metal core. But what could be the conductor on Europa, which is so receptive to the Jovian magnetic field? It turned out to be liquid water.

The smooth outer shell of Europa was the first clue that made scientists consider whether unseen reserves of liquid water might play an important role in the surface geology. *Galileo*'s magnetometer measurements allowed scientists an unprecedented view beneath the icy crust and revealed the presence of a lunar-wide ocean some one hundred kilometers deep! Even further, scientists knew that pure water would be a relatively poor conductor of electricity and that the ocean must be salty at some level to produce the observed magnetic field. Just how salty the ocean might be depends on the type of minerals that are considered. But there is good reason to believe that the Europan ocean is just as salty as Earth's own, or even more so.

How much liquid water are we talking about on Europa? If the Europan ocean is truly as deep as the Galileo magnetometer measurements suggest, then the total liquid volume is approximately two times as large as that of Earth. It is an astonishing realization that this one moon, a shade smaller than Earth's moon, contains more liquid water than our entire planet. If any one discovery tells us to sit up and take notice, it is that this ice moon and its cousins in orbit around Jupiter and Saturn quite likely contain the bulk of the liquid water in the solar system.

Before we even begin to comprehend what this might mean for life to exist within these moons, I think you should be allowed the opportunity to call a halt and ask for a big sanity check. We know that the surface of Europa is made of water—we can tell that by observing the clear spectroscopic signature of water absorption in the spectrum of reflected sunlight. However, the average surface temperature of Europa is 110 Kelvin (−160° Celsius) at the equator and a truly frigid 50 Kelvin (−220° Celsius) at the poles. Just what on Europa is allowing liquid water to form—remember that we need a temperature between 273 and 373 Kelvin (0° to 100° Celsius) under "normal" pressure? Where is the energy coming from to produce this heat in the frigid depths of the outer solar system?

Tides, Resonance, and Energy

In an age when we are rightly concerned with harnessing renewable energy resources, have you ever paused to consider where tidal energy comes from? We experience ocean tides because the combined gravitational force on Earth exerted by the sun and the moon is slightly greater on the side of our planet facing these bodies than on the side facing away. This in turn deforms Earth, to a tiny extent, from a sphere to a rugby ball shape. Earth's oceans, being more flexible than rock, bulge to a slightly greater extent. The two high tides we experience each day occur when Earth's rotation carries us underneath the near- and farside ocean bulges that more or less always face the combined sun-moon direction.

The energy that causes the oceans to form this tidal bulge is gravitational—it comes from Earth's orbit about the sun and from the moon's about Earth. As a result, the orbital radii of both Earth about the sun and the moon about Earth are increasing with time at an almost infinitesimally slow rate. So the electrical energy used in your home and generated by your local tidal barrage (should you live near the Rance estuary in France, say) is ultimately drawn from the orbits of Earth and the moon.

But what of Jupiter and the Galilean satellites? In this case, it is exactly this same effect: tidal energy provides heat in the interior of each moon. Io and Europa—and to a lesser extent, Ganymede and Callisto—are deformed by Jupiter's gravitational field. The critical difference from the Earth-moon tidal interaction is that the orbits of the Galilean satellites are much more elliptical than the moon's, with the result that at some points in their orbit they are closer to Jupiter and at others they are farther away. This causes the tidal deformation to vary within each lunar orbit—the surface of Io buckles and bends by as much as one hundred meters during its forty-two-hour orbit about Jupiter.

To imagine the effect of this, think of a squash ball. Before each game, the players spend a few moments squashing and squeezing the ball to increase its flexibility and bounce. The energy expended as you deform the ball creates heat arising from friction between the molecules that make up the rubber ball. In the same way, the cyclical deformation of the inner moons of Jupiter grinds away at each lunar interior, producing heat from the resulting friction in the rocky core. But there is one additional effect of critical importance to maintaining this tidal effect.

In a single planet-moon system, the orbit of the tidally deformed moon will become circular as the gravitational interactions iron out wrinkles in the orbit. But the inner moons of Jupiter have entered into a tidal resonance in which their gravitational interactions have synchronized their orbits—with the result that for every four orbits completed by Io, Europa completes two, and Ganymede completes one.[4] When the inner moons periodically align, the system receives a small gravitational kick that maintains their elliptical orbits.

Where does the energy come from? Just as the tidal effects we experience on Earth have changed the rotation of Earth and the orbit of the moon, the same occurs with Jupiter: Jupiter rotates more slowly than it otherwise would, and the orbital radii of the inner moons are becoming (very, very) slowly greater with time. Where does the energy go? Directly into the interior of each moon, providing what can be formidable sources of heat: day in, day out, as much heat per square meter of its surface is produced by Io as flows through Yellowstone National Park. It is a lunar-wide volcanic hot spot.

Tidal energy stokes this fiery cauldron just as it warms the interiors of Europa and Ganymede. The diminishing activity of each of the inner moons of Jupiter can be understood from the mathematics of their tidal interactions: the tidal forces experienced by each moon decrease as its distance from Jupiter increases. Each moon in turn experiences less tidal heating, and each lunar surface is disturbed to a lesser and lesser degree by the interior ructions thus produced. We have seen that this heat produces lunar-scale volcanism on Io—but what might we expect on Europa? The thermal energy provided by tidal heating on Europa appears to be enough to warm a lunar ocean. Only at the outermost layers, farthest from the warm core, does the howling cold of deep space encase the moon in a thick sheen of hard ice. But if tidal heat produces violent volcanism on Io, should we expect to discover milder versions on Europa? Could there be undersea volcanoes on Europa—each powered by the gravitational grasp of Jupiter—and what would this mean for the possibility of life within this distant moon?

Fire in the Abyss

On the night of February 15, 1977, scientists aboard the research vessel *Knorr* were sailing in the Pacific Ocean roughly midway between the Galapagos Islands and the coast of Ecuador. They were about to make one of the most important discoveries ever made about life on Earth. Below them, at a depth of 2.5 kilometers, an unmanned deep-sea probe—ANGUS (Acoustically Navigated Geological Undersea Surveyor)—was being carefully towed just above the surface of the sea floor.

The geologists and oceanographers on board that night knew what they were looking for—a deep-sea hydrothermal vent, where ocean water, circulating through Earth's crust, is heated by the underlying mantle and forced to the surface in columns of superheated water. Although they had long been suspected of being associated with volcanically active midocean ridges, no one had ever discovered more than circumstantial clues to their existence. Late in the evening, the scientists monitoring the probe noticed an anomalous yet significant temperature spike in the waters of this abyssal deep. With the first rosy fingers of dawn in the air, ANGUS was laboriously drawn to the surface, and the series of three thousand photographs taken during the sixteen-kilometer night cruise was developed and studied. Bob Ballard, the veteran oceanographer, takes up the story in an article written at the time: "The photograph taken just seconds before the temperature anomaly showed only barren, fresh-looking lava terrain. But for thirteen frames (the length of the anomaly) the lava flow was covered with hundreds of white clams and brown mussel shells. This dense accumulation, never seen before in the deep sea, quickly appeared through a cloud of misty blue water and then disappeared from view. For the remaining 1,500 pictures, the bottom was once again barren of life."

Over the next two days, members of the scientific team jostled for spots in *Alvin*, the expedition's deep-sea submersible, to witness the scene with their own eyes. Through cracks in the broken sea floor, turbulent currents of hot misty water bubbled to the surface. Strewn about these flowing columns of water was an unthought-of abundance of deep-sea life, clearly thriving in what scientific orthodoxy had until then considered the sterile depths of the ocean. Not surprisingly, at this point the expedition members really regretted not having invited a biologist along. It took another two years for a biological expedition to return to this deep-sea hydrothermal vent. It was there, confronted by a deeply alien yet ultimately terrestrial environment, that biologists began to realize just how important these unique ecosystems would prove to be. Holger Jannasch, a member of that 1979 expedition, was among the first scientists to arrive at the key realization: "We were struck by the thought, and its fundamental implications, that here solar energy, which is so prevalent in running life on our planet, appears to be largely replaced by terrestrial energy—chemolithoautotrophic bacteria taking over the role of green plants. This was a powerful new concept and, in my mind, one of the major biological discoveries of the 20th century."

Extremophile microbes living within the rocky walls of the scalding vents were thriving on chemical reactions that extracted energy—geochemical rather than solar—from the minerals dissolved within the superheated water. Tiny shrimps, giant clams, and oversize blood-red tubeworms in turn fed on this microbial feast while larger predators, crabs and small fishes, prowled the periphery—all in the total absence of sunlight. The fragile cord linking life on Earth with solar energy had been well and truly severed by this thriving, alien ecosystem—and astrobiology would never be the same again.

Astonished scientists lost no time in casting their gazes back to the icy moons of Jupiter, where the salty oceans of Europa immediately took on a new context. The presence of dissolved mineral salts in the oceans of Europa meant that the water had to be in contact with the rocky lunar interior. The all-powerful second law of thermodynamics demands that tidal heat produced within the core of Europa diffuse outward to the cooler surface. The discovery of hydrothermal vents provided solar system scientists with a compelling vision of what this geological interface between rock and ocean might look like. Could deep-sea biological systems similar to ones now known on Earth thrive within Europan hydrothermal vents? What previously unimagined denizens of the deep could be brought forth to populate these distant lunar oceans?

It is difficult to overstate the extent to which the discovery of deep-sea hydrothermal vents on Earth has transformed our view of the habitability of the outer solar system. However, I want to dwell a little while longer in these fantastic realms and celebrate some further discoveries with you.

In 1979, tall chimneys of deposited minerals were discovered at latitude 21° north near the Gulf of California. Once again, scientists in *Alvin* were astonished to see clouds of jet-black water belching violently from these chimneys, like smokestacks of the very darkest industrial ages. Dissolved minerals in the superheated water produced the black color when they precipitated out of solution after making sudden contact with near-freezing ocean water. When scientists inserted the first temperature probe into the chimneys, it promptly melted;[5] only with a hastily reengineered probe did the expedition team record a water temperature of 350° Celsius. Imagine, then, the astonishment of later scientists who in 1997 extracted the wonderfully named archaean microbe *Pyrolobus fumarii* (the fire-lobe of the chimney) from within another such a hydrothermal vent—metabolizing away at a balmy 113° Celsius. The largest such chimneys are now known to top out at some sixty-one meters tall—each represents a new monolith challenging our previously held ideas of the origin of life on Earth and the prospects for life beyond it.

Trapped under Ice

Imagining an abundance of alien life in the oceans of Europa is one thing. Getting at it, under an ice sheet that may be thirty kilometers thick is—to put it mildly—quite another. But there are environments on Earth that can teach us how we might go about it.

Have you ever flown a radar-equipped Hercules C-130 over the vastness of Antarctica? Nope, me neither. I imagine that, like many examples of scientific data gathering, it involves moments of great beauty interspersed with long stretches of interminable boredom. The mid-1970s witnessed one such heroic act of data gathering when hundreds of research flights traveled some four hundred thousand kilometers back and forth across the Antarctic. On each journey, underwing radar arrays produced a downward-facing beam of 60 MHz radio energy.[6] The results formed a wonderful three-dimensional map of the continent; the radio beams had penetrated deep into the ice sheet to reveal its interior structure.

Cold ice is relatively transparent to radio waves—except at interfaces in the ice layers, where a small amount of radio energy is reflected back to the surface. Careful signal timing can reveal the depth of each interface. Ancient strata in the ice appear as rippled layers akin to the warped growth rings of trees. The rocky base of the continent forms an indistinct, undulating layer. Subglacial lakes—large bodies of fresh water trapped at the base of the ice sheet—return a strong, mirror-flat radio echo.

Such lakes were not unexpected: several kilometers of ice overhead create unimaginable pressure, which lowers the temperature at which water can exist as a liquid. Where the pressure-affected melting point of water dips below the deep-ice temperature, liquid water can form. Then, if the local topology of the bedrock forms a trough or a hollow, that water can persist and build up into a sizable lake. Just how big is shown by Lake Vostok, which lies some four kilometers beneath the East Antarctic Ice Sheet.

Lake Vostok covers an area of some fifteen thousand square kilometers and has a typical depth of over four hundred meters. Though the largest by far, Vostok is only one of many subglacial lakes to have been discovered dotting the depths of the Antarctic landscape. In many ways, these lakes represent stark, alien worlds. The overlying ice shows layered strata stretching back over four hundred thousand years of history. But since the overlying ice gradually flows down to the ocean, and since strata are renewed, the lakes may have been isolated from the outside world for much, much longer—up to twenty-five million years. That represents a truly interesting challenge for life—no sunlight, bare rock, and only the slight possibility that the flow of subglacial water might circulate eroded minerals.

In many ways, the challenge is similar to that offered by Europa. Both are icy worlds with the potential to host unknown yet potentially unique ecosystems. Even on Earth, however, it is no easy feat to drill through as much as four kilometers of ice and perform an environmentally sensitive study of subglacial waters that may have lain undisturbed for eons.[7] Care must be taken at every step, yet time is of the essence. Precariously camped out on the exposed polar plain, you and your team must melt your way down with either a hot-water probe or an antifreeze-equipped rotating drill bit. Precious fuel has been flown in, and once you have drilled down to the icy depths, you may have only enough fuel to keep the rapidly refreezing borehole open for forty-eight hours at most.

Dedicated scientists have so far made the delicate and demanding breakthrough at two such lakes: a U.S. team at Lake Whillans and a Russian team at Lake Vostok. What did they discover? Dark murky worlds potentially rich in biological material—both floating free in the lake water and within the deep, undisturbed sludge of the lake bottom. First contact with these deep lakes occurred in late 2012 and early 2013, and it is too early to yet be sure of the detailed biochemical nature of any life discovered within them. Their discoveries may prove to be as profound as those revealed by the worlds of the deep-sea hydrothermal vents. We just don't know yet—and you should definitely stay tuned.

But before you allow your dreams of deep-ice drilling to leap from Antarctica to Europa, I have to weigh you down with the heavy burden of our current ignorance: we don't even know how thick the ice is on Europa. *Galileo* did not carry the kind of ice-penetrating radar used to map the subglacial worlds of Antarctica. Europa's outer crust of ice could be a few kilometers thick; it could also be thirty kilometers thick. We remain ignorant of the dynamics of the ice, too. Is it recycled via a cryogenic version of plate tectonics? Are observed cracks in the ice evidence of faults where liquid water might approach the surface? Could there be water reservoirs closer to the surface, within the ice layer itself?

Despite the exciting prospects for life that Europa appears to offer, there is much—too much—that we do not know. To answer these questions, we must return to Europa with a new probe, one with new instruments for answering these questions.

At least one space mission has already purchased its ticket and will return to Europa around 2030. *JUICE* (*Jupiter Icy Moon Explorer*) has been selected (and more important, funded) by the ESA as its next large solar-system-exploration mission. The price tag is 900 million euros, and the project is already at the advanced planning stage. Is this the right mission to unveil the secrets of Europa's icy crust and discover new life within the solar system? Er, probably not. To be fair, space scientists are rightly measured in their approach to discovery, and *JUICE* will spend its time obtaining detailed images of the surfaces of Jupiter's moons and using its ice-penetrating radar to look within their frozen crusts. The main target of the *JUICE* mission is Ganymede rather than Europa—which will receive just two flybys in the Ganymede-approach phase. One of the reasons for this focus is practical: it requires less spacecraft propellant to enter Ganymede's orbit than Europa's. Since propellant equals liftoff mass, and mass equals money, this is a real limit.

Is anyone planning to return to Europa, an icy moon that we know to harbor a warm liquid-water ocean? NASA is considering the Europa Clipper—a multiple-flyby mission to the Jupiter-Europa system that would encounter Europa up to thirty-two times. Like *JUICE*, it would carry high-resolution cameras and an ice-penetrating radar with which to map both the surface and the depths of the Europan ice shell. Why not a Europa orbiter with a simple lander to test the composition of the surface ice and discover just what is that pinky-orangey-brown stuff? The answer is a general lack of crinkly green stuff—money. Cash for new missions is painfully tight because of NASA's plans for (yet another) Mars rover. A stripped-down Europa Clipper mission still tips the budgetary scales at some $2 billion. If you want a Europa orbiter with a simple lander, you could be looking at twice that price. What you, the national space agencies and the communities of scientists who stand with you, have to decide is whether Europa (as a case in point) is a high enough priority to push other ideas and questions off the table.

Galileo's *Fiery Death*

While Galileo Galilei escaped burning at the stake for his astronomical heresies, the space probe that bore his name was not so fortunate. In 2003, having performed what remains to this day the definitive observations of the inner moons of Jupiter, the *Galileo* space probe was deliberately flown into the atmosphere of Jupiter at a staggering 174,000 kilometers per hour. Unseen from Earth, it made its final, fiery descent into oblivion, sharing the fate undergone by its descent probe eight years earlier.

Why was *Galileo* condemned to this intentional destruction? The answer gets to the heart of human contamination of the moons and planets of the solar system and our efforts to protect the potentially pristine habitats for life that they provide. The *Galileo* probe was not sterilized before leaving Earth, so it carried a cargo of terrestrial bacterial voyagers to the outer solar system. Once we discovered that Europa harbors a vast lunar ocean, it would have been a great blunder to allow the fuel-depleted *Galileo* probe to potentially crash onto the surface and deposit a shaken, stirred, but otherwise living host of bacterial pilgrims on the new world of Europa.

Many bacteria might perish on arrival as they adapted to the new, harsh environment (though no more harsh than the frying pan from which they had just been delivered). But any survivors that gained access to the Europan ocean could then float freely throughout what may be a precarious, pristine environment—forever changing the biological landscape before we could launch a dedicated search for life. In the end, although the risk was deemed to be small, the consequences were known to be great, and the aging Galileo was consigned to a terminal trajectory. Inquisitive to the end, *Galileo* performed a de-orbit maneuver that carried it past Jupiter's moon Amalthea, allowing for a precise measurement of the mass of this poorly known moon, before it made a graceful farewell arc.

Such bacterial stowaways are an example of what we call forward contamination, and our experience with *Galileo* demonstrates why they should be avoided—it would be painfully ironic if the first life detected beyond planet Earth was a delinquent terrestrial hitchhiker.[8]

You may well ask whether forward contamination is not a greater risk for Mars, given the large number of probes dispatched to its surface. Present-day Mars spacecraft are subjected to a pretty thorough prelaunch scrub down intended to leave them spiffy and, hopefully, bacteria free before their journey to the red planet. Is this necessary, given the large number of unsterilized craft sent in the past, as well as the likelihood that over billions of years, large amounts of bacteria-bearing Earth rocks were blasted into space by meteor impacts and landed uninvited on Mars? For the specific case of Mars exploration, the vocal minority may well have a point. But given humanity's long history of leaving muddy footprints[9] over newly discovered lands, perhaps we should be forgiven for once for exercising more care than required.

Having come up to speed on forward contamination, we don't have to make a great leap of insight to realize that back contamination of Earth—bringing alien life back home to visit—also gives NASA's planetary protection officers many a sleepless night. We have returned scientifically sampled solar system material to Earth four times: lunar samples returned by U.S. manned and Soviet unmanned sample return missions, samples of comet Wild 2 returned by the *Stardust* probe, and samples of near-Earth asteroid 25143 Itokawa returned by the *Hayabusa* mission.

Attempts to protect Earth from these samples have varied in their approach: rock samples from the Apollo missions were stored in triple-sealed containers for their journey from the moon. The *Stardust* probe parachuted down near a U.S. army base and was transported to the Johnson Space Center in a secret operation eerily reminiscent of something out of *The Andromeda Strain*. The *Hayabusa* sample return capsule was put in two plastic bags filled with nitrogen. In some sense, these examples miss the point because they were not sampling environments where any kind of biological activity was expected.[10]

But what about the next generation of sample return missions—from Mars and the moons of Jupiter and Saturn—which we are excited about exactly because they may be visits to biologically active environments? The first mission to return potential biological samples from another world in the solar system faces a punitive budgetary hurdle: no one will allow anything to be brought back to Earth unless it is delivered to a dedicated sample-handling facility operating under the very highest standards of biosafety.

In fact, given the surprises that Earth bacteria continually spring on us (turning up in our supposedly sterile spacecraft clean rooms, for example), we don't really know even how safe we should plan to be. Is it sufficient for the samples to make an airtight journey from the spacecraft to a receiving facility—or is an even higher level of quarantine required? Creating such a facility from scratch always looms large on any list of mission budget items. Too important to ignore yet too expensive to fund (at least within the frugal environment of a single mission—even if it is a large one).

The solution is not difficult to arrive at. How many sample return facilities would we likely need, given the scope of our ambitions for solar system exploration? Probably one. Should the specifics of whether the sample has been collected from Mars or Europa affect the overall design of the facility? Probably not. Should national space agencies sit down and consider collaborating on building such a facility? I think you know what my answer might well be.

You have probably realized that if you have a secure sample return facility on Earth, you still need a foolproof method of delivering your sample to the door (that means from low-Earth orbit through reentry and controlled landing on Earth). Given the care—and thus the strict rules—with which we transport humans from Earth's surface to low-Earth orbit and back again, should we return appropriately packaged samples to Earth under human-spaceflight requirements? What level of risk is appropriate? Once again, this idea calls out for space agencies to put their heads (and wallets) together to consider whether a collective approach will provide the next generation of sample return missions the critical Earth infrastructure required.

The Miracle of Enceladus

Perhaps I have been too hard on you in this chapter. Europa offers real potential as a habitat for life beyond Earth. Abundant in liquid water, powered by tidal energy, and as rich in organic chemicals as any location in the solar system. But then I dashed your hopes by detailing the hard challenges of actually getting there, penetrating the ice, and revealing the alien ecosystems within—not to mention the ever-present hurdle of paying for it all. At this point, therefore, I want to cheer you up—and I am going to do it with Enceladus.

Enceladus is a tiny moon—one-sixth as large as Europa and one-quarter of 1 percent as massive—whose surface is pure water ice, just like its larger Jovian cousin's. Enceladus orbits relatively close to Saturn, about twice as far from the planet as the outer edge of the main rings, familiar to all keen observers of the night sky. The miracle referred to in the section title—you could choose "extreme surprise" if you prefer—is that a moon so small is so active.

In 2005, to their surprise, astronomers inspecting the latest *Cassini* images observed geysers of material spouting from a localized region near the south pole of Enceladus. *Cassini* has even flown through the geyser plumes in daring flyby encounters and sampled the ejected material with its onboard mass spectrometer. The geysers are composed of salty water ice and water vapor—good old-fashioned sodium chloride—and an unknown assortment of likely organic compounds. So is this tiny ice moon like Europa, a potential water world with a subsurface ocean? The answer seems to be yes. The icy surface of Enceladus shows both old, heavily cratered regions and young, unblemished terrain. The tidal heating it experiences as it orbits Saturn is similar in amount to that undergone by Europa, and tiny variations in *Cassini*'s flyby trajectories about the moon are consistent with the presence of a ten-kilometer-deep regional ocean lying beneath the ice at the South Pole.

The continued successful operation of the *Cassini* probe—well beyond its initial mission, which expired in 2008—has allowed studies of the geyser regions of Enceladus in exceptional detail. In 2014, the results of these dedicated observations confirmed the existence of 101 individual geysers, each located at a local hot spot in a series of surface faults, known as the "Tiger Stripes," close to the southern pole. The faults appear to be due to the cracking of the ice sheet as tidal forces deform Enceladus during each thirty-one-hour orbit of Saturn. The source of the heat in the Tiger Stripes is a compelling clue in itself—rather than being escaping tidal heat, it appears to come from the energy released as some of the liquid water, welling up from within the moon, freezes at the mouth of the surface cracks. The rest—about two hundred kilograms of material per second—escapes as water vapor and ice crystals and makes up the diffuse yet giant E ring of Saturn.

So there you have it—an ice moon with a subsurface liquid-water ocean, rich in salt and organic chemicals. With rare and fortuitous exuberance Enceladus gently spouts this material into space for any passing astrobiologist to sample and return to Earth—where the secret inner world of this icy moon can be revealed in exquisite laboratory detail. It is most definitely not too good to be true. The opportunity provided by these ethereal geysers to swim vicariously the subsurface oceans of Enceladus is both genuine and too compelling to miss.

Follow the Plume

How will we do it—travel to Saturn, encounter Enceladus, capture particles from the wispy geysers, and return them safely to Earth? How ambitious or even realistic is such an idea? Though it is indeed a bold plan, you should certainly be encouraged by the fact that we have already successfully accomplished each element of the potential Enceladus mission: multiple planetary gravity assists in the inner solar system have provided free rides for *Galileo* and *Cassini* to Jupiter and Saturn, respectively—boosting spacecraft cruise velocities and reducing travel times to the outer planets.

Once the spacecraft encounters Enceladus, aerogels—low-density solids that are largely empty space, developed for the *Stardust* and *Hayabusa* space probes—could collect icy particles and gases from the geysers. Though the geysers are visually impressive, it can be surprising just how tenuous the streams of particles are—typically there is just one microscopic grain of material per cubic meter at a height of some eighty kilometers above the surface of Enceladus. Multiple flyby trajectories would be required to sample enough geyser particles, perhaps drawn from locations across the southern pole.

Although both *Stardust* and *Hayabusa* have successfully returned samples to Earth, no craft has done so at the higher velocities that would occur when returning from Saturn—and certainly not with such a potentially contentious, biologically active cargo on board. *Stardust* returned to Earth at a velocity of just over six kilometers per second, *Hayabusa* at twelve. An Enceladus mission would return at a velocity of sixteen to eighteen kilometers per second. Since kinetic energy increases with the square of velocity, the Enceladus spacecraft would have to safely burn up between two and nine times as much energy on reentry as *Stardust* or *Hayabusa*.

The complete package is presented as the Low Cost Enceladus Sample Return Mission.[11] Launch would occur in 2021, and arrival at Saturn eight years later would follow the gravitational VEEGA trail blazed by *Galileo* thirty years earlier. The probe would spend two years in the vicinity of Saturn, carefully tuning its orbit in a series of gentle swoops to glide through the geysers. The return journey, rolling down the gravitational hill toward the sun, would require four and half years to bring the extraordinary samples to Earth—finally returning in 2037.

Can we do it? Yes, we can! Wait a minute—how much will this gripping endeavor cost me? At present the Low Cost Enceladus Sample Return Mission is being prepared as a NASA Discovery-class mission, and that means a price tag of around $500 million. But that budget does not cover a couple of items on the shopping list: NASA would need to chip in a launch vehicle plus a power source capable of operating for over seventeen years in deep space—a challenging, long-term goal.

And finally there is the sample return challenge, both delivery from low-Earth orbit and curation on Earth. The Enceladus mission designers have investigated whether the velocity at which the geyser particles hit the aerogel might act to presterilize the sample—the impact energy could be tuned to break apart larger, biologically active compounds while retaining their chemical constituents for study on Earth. This would of course mitigate against the need to adopt strict sample return protocols. But in many ways this misses the point: if multiple space agencies are interested in the biological exploration of the solar system (which they clearly are), then at some point they are going to have to chip in to create a sample return facility. One can only hope that an Enceladus sample return mission can motivate them to face up to this important and necessary decision.

I have to admit: it all sounds pretty compelling and eminently achievable. Would an Enceladus sample return mission be my highest priority? Well, it's certainly up there (though there are enough revelations still to come that I am not going to commit just yet). Whatever we choose to do, I hope that the next time you gaze at Jupiter and Saturn—whether with the naked eye or with telescope in hand—you savor the knowledge that within their retinue of moons lie oceans of warm salty water. Oceans that are not, in the grand scheme of things, all that different from the warm salty water that constitutes the bulk of our cells and that we retain as a chemical memory of our origins. Whether the ice moons of Jupiter and Saturn have their own biological story to tell remains for us to answer. But with their discovery alone, these moons and their oceans present a stunning reality in which the solar system beyond Earth provides a richer range of habitats for life than we could have dared to imagine.

Why in our story of the search for life in the universe does Titan get a chapter all to itself? Come to think of it, at this point in the book you may be wondering—with some justification—why we are still within the solar system. When are we going to move on to new star systems, new planets, new life? Please tell me this isn't going to be one of those "life in the universe" books that just drones on about the solar system and leaves out all the neat stuff that I am interested in. No, it isn't. But thank you for your patience in sticking with me so far.

So why Titan? Why does it make the top five? The quickest answer is that it checks all our boxes for a potential habitat for life—with one critical difference. It is a planet-sized moon—the first to be discovered orbiting Saturn—larger than Mercury and only a shade smaller than Ganymede. It has an atmosphere—thicker than Earth's but neither as hot nor as toxic as that of Venus. To say that it is rich in organic chemicals is the biggest understatement in this book—it is the solar system's very own petrochemical plant, producing an uncountable multitude of complex organic chemicals. It even has bodies of stable liquid—lakes and rivers—not tucked away under an impenetrable ice sheet but basking on the surface.

So what's the catch? It's cold. Very cold: 90 Kelvin, or −180° Celsius. Too cold for liquid water but—in the best tradition of an alien Goldilocks—just right for liquid methane and ethane. That is the critical difference. All the water-based chemistry on which Earth life is based is profoundly ill suited to Titan's seas of organic solvents. Yet all the elements of life are present: liquids, energy, and organics. Should life exist within this unfamiliar environment, it will be truly different from life on Earth—chemically alien. That is why Titan matters and why I want to take you there.

Through an Anagram Darkly

Christiaan Huygens discovered Titan in 1655, some forty-five years after Galileo discovered the inner moons of Jupiter. Huygens used the same discovery method pioneered by Galileo: repeated telescope observations that tracked the moon in its orbit—about Saturn in this case—and led him to estimate the period as sixteen days and four hours (the modern value is only six hours shorter). Both Christiaan Huygens and his brother Constantijn were accomplished lens grinders, and their ten-foot-long telescope yielded an impressive magnification of 50×—five times as powerful as that used by Galileo to observe Jupiter.

The method by which Huygens announced his discovery reveals a neat quirk of seventeenth-century science. In the summer of 1655, he circulated an anagram to his colleagues and peers that read *Admovere oculis distantia sidera nostris vvvvvvv ccc rr h n b q x*.[1] Only in the following year—when he was presumably more certain of his discovery—did he publish a pamphlet in which he explained the anagram as encoding *Saturno luna sua circunducitur diebus sexdecim horis quatuor*, which, as those of you with a fine classical education know, means *Saturn's moon revolves every sixteen days four hours*. Unfortunately, scientists no longer circulate anagrams to their peers in order to claim primacy in new discoveries. Scientific preprint servers—which authors use to present submitted yet unpublished results—represent an admittedly effective yet ultimately unromantic modern alternative.

A Hazy Shade of Titan

Before the flyby of *Voyager 1* on November 12, 1979, Titan was essentially an orange smog-bound moon with no discernible surface features. Even after *Voyager 1* left, it remained an orange smog-bound moon with no discernible surface features. *Voyager* did, however, reveal rich details of that dense, smoggy atmosphere.

The atmosphere of Titan is almost entirely composed of nitrogen—95 percent by mass—in the form of N_2. The rest is almost entirely methane—CH_4. It is Titan's trace chemistry that is so fascinating: molecular hydrogen, H_2, and a bustling horde of hydrocarbons from the very simplest—which we can recognize from their distinctive spectroscopic emission lines—to an unrecognizable mass of organic molecules whose complex spectral properties defy disentanglement.

Voyager revealed that Titan's atmosphere is hazy rather than cloudy. The high-altitude haze layers are just like earthbound smog. Much of this smog is made of microscopic particles—dubbed tholins by Carl Sagan and Bishun Khare[2]—each composed of solid organic molecules that are light enough to be suspended high in the atmosphere.

The most striking aspect of Titan's atmosphere is how much of it there is. Overall, the atmosphere is about 20 percent more massive than Earth's. But the surface gravity on Titan is only about 14 percent of that on Earth, and as a result the atmosphere is tenuous and extended. The combination of atmospheric mass and surface gravity generates a surface pressure on Titan that is about 1.5 times that of Earth's. Yes, you could walk around on the surface of Titan without a space suit—though you would require warm clothing and an oxygen mask!

The astute among you may question how Titan has kept its atmosphere in face of the stiff wind of energetic, electrically charged solar particles. This is indeed a good question—Titan has no magnetic field of its own. But for most of each orbit it lies within the magnetic field of its parent, Saturn, and is thus shielded from the ablative solar wind. An additional possibility is that Titan's atmosphere is continually replenished via low-temperature volcanism—cryo-volcanism—which vents methane and other gases to the surface.

What powers the fascinating—and to a large extent, unknown—chemistry of Titan's atmosphere? The sun. Titan is an active photochemical environment. Though fewer photons strike each square meter of the top of Titan's atmosphere than hit Earth's, the typical energy of each photon is no different.[3] Solar photons thus bombard the upper atmosphere of Titan with sufficient energy to break apart—photodissociate—methane. The broken fragments then recombine with nitrogen, hydrogen, and the rest of the atmosphere in an organic chemical cascade.

A Strange New World

Only when the *Cassini* spacecraft reached Titan in 2004 and dispatched the *Huygens* lander to the surface did we truly realize what a strange, compelling world we faced on Titan.

The *Huygens* probe—a small static lander—parachuted to the surface of Titan, touching down on January 14, 2005. During the descent, *Huygens* used its unique vantage point to image the previously unseen surface. It witnessed low pale hills of rock-hard water ice, riven with myriad dark river-like channels. *Huygens* did not so much touch down as squelch down onto the surface.[4] There it captured a handful of enigmatic images of the surface: a plain of dark organic mud peppered with small pebbles and rocks of water ice. Although *Huygens*'s batteries survived the deep cold of the surface for only ninety minutes, the mission represented a breakthrough in solar system science—this was the first time we had traversed the asteroid belt to land on either a planet or a moon in the outer solar system.

Over the past decade, the *Cassini* probe has performed more than one hundred flybys of Titan. Although the haze layer absorbs most optical and infrared radiation, it is largely transparent to radio waves. By exploiting this effect, *Cassini*'s radar instrument has provided unobscured views of the surface of Titan. Rather like mowing stripes on a lawn, each flyby encounter allows *Cassini* to image only a narrow strip of surface terrain, but even with this limitation, *Cassini* has imaged about 50 percent of the surface of Titan. These radar images reveal that large contiguous areas of the surface present an almost mirror-like flatness. As we saw with radar images of the Antarctic ice sheet, this is the signal of liquid—but in this case, rather than water, the temperature and composition of the atmosphere point the finger at methane and ethane.

In addition, the radar data have provided an unprecedented glimpse within the lakes themselves: the subtle attenuation of the radar signal as it penetrates the liquid layers—a technique known as radar bathymetry—has shown that the lakes of Titan vary considerably in their character. Some are little more than shallow wetlands, on average a few tens of centimeters to a few meters deep. Others, such as Ligeia Mare, are as large as the Great Lakes of North America, with depths of 170 meters.[5] Finally, in 2009, the lakes revealed themselves in spectacular fashion to *Cassini*'s visible and infrared mapping spectrometer when it caught the flash of sunlight reflected from their surfaces. Such specular reflections—exactly the same as the glare of sunlight reflected from Earth's lakes and oceans—provides unambiguous evidence of large bodies of liquid on the surface of Titan.

After the surprise discovery of lakes came new radar images of rivers of methane and ethane, as well as wide deserts apparently covered by high dunes of organic sand. The presence of rivers—drainage channels—chimed with an analysis of the temperature and pressure of Titan's atmosphere, which predicted that methane would evaporate from lakes to enter the humid lower atmosphere and eventually fall as rain on the surface. In other words, Titan has a full atmospheric methane cycle, which mirrors the hydrological cycle on Earth.

Titan seemed to be a world like no other. But the more that scientists thought about Titan, the more they realized that if it resembled any world in the solar system, it seemed closest to Earth in its chemical composition and physical processes. But not the Earth of today. Instead, Titan resembled the early Earth before life arose—the world of Miller-Urey, where oxygen was absent from the atmosphere and complex organic chemicals were readily synthesized. That is a pretty provocative statement, though—if you follow it to its conclusion. Though our knowledge of the origin of life on Earth is incomplete, our current belief is that life arose naturally from the active chemical environment present in early times. Could Titan today mirror those ancient chemical conditions? If so, might it be a top contender to host life? For some, that suggestion is too provocative. Many have cited the absence of liquid water—so familiar to Earth life—and the low temperatures (compared to Earth's) as fundamental barriers to biochemistry.

But these differences and their attendant challenges to our Earth-based view of biochemistry are precisely the reasons that make Titan so interesting. Titan compels us to think differently, to question previously accepted truths. From Titan, as we look back to consider life on Earth, we are forced to face our preconceptions. And that can only prove to be a good thing.

Rare Titan

In 2000, Peter Ward and Donald Brownlee published a groundbreaking book called *Rare Earth*. In it they considered many of the central ideas of astrobiology, but also the quirks, fine-tunings, and coincidences that lay along the circuitous path to the emergence of complex life on planet Earth. Their major conclusion was that planets exactly like Earth, populated by creatures exactly like you, might indeed be rare. Their minor conclusion was that Earth might well be only one among a plurality of inhabited worlds populated by a wide diversity of basic life. But in the minds of many readers, these conclusions have become incorrectly conflated to read, "Earth is special, life is rare."

Chris McKay, an astronomer at the NASA–Ames Research Center, has played on some of these misconceptions with the idea of Rare Titan. McKay contemplates a Titanian astrobiologist pondering the possibility of life on Earth. The imaginary Titanian astrobiologist realizes that this strange blue-green alien planet with its abundant liquid-water oceans and oxygen-rich atmosphere would be completely toxic to all the biochemistry on which Titanian life is based. Even more, at the high surface temperatures encountered on Earth, many of Titanian life's basic molecules would rupture and break. Earth therefore would represent a poor target for Titan's next astrobiology mission.[6] The idea of this playful daydream is that we will have to drop some of our emotional baggage regarding the necessity of water and moderate temperatures for life if we are going to be open-minded enough to both search for and ultimately recognize even mildly alien life.

Water, the Enema of Life

So how might we lighten our load of preconceptions in order to prepare ourselves for an astrobiological journey to Titan? Well, let's start with water.

Many biologists will queue up all day long to tell you how important water is to life on Earth. But the important question is whether water is *the* liquid for life or just one possible liquid among many that could serve as a medium for biochemistry. At present we don't know—mainly because we don't even have a complete understanding of Earth life, let alone life in general. To be fair, many astrobiologists recognize this, and thus the rallying cry "Follow the water!" is often accompanied by the qualification that water is the liquid used by life on Earth—the only life we currently know. So for this chapter, I would like to rebel against this view. Instead, I will tell you about some of the ways in which water is an impediment to life and how life has either adapted to this impediment or just sucked it up and gotten on with the process of living.

Water is the archetypal polar liquid—though in this case, polar is a chemical rather than a geographic term. Put hydrogen in a molecule with a strongly electron-attracting atom like oxygen, and although all the atoms share their electrons (and thus form a covalent bond), some share better than others. Oxygen takes the lion's share of the electrons, and as a result a small charge imbalance, an electric dipole, exists across the molecule.

Where such opposite polar charges attract one another, they form weak hydrogen bonds—which can form between or even within polar molecules. Hydrogen bonds help other polar molecules—like salts, proteins, even DNA—dissolve in water and thus gain the freedom to move and react with other dissolved chemicals. In this way, water can be said to be "good" for life—since life on Earth uses a heady mix of salts and proteins in its chemistry.

But there is a dark side to the relationship between water and life. The nucleobases (C, A, G, and T) that make up our DNA show a worrying tendency to break down in water—polar molecules can disrupt weak covalent bonds via a reaction referred to as hydrolysis. As a result, DNA needs to be regularly repaired in order to preserve the integrity of its genetic code. The polar nature of water also impedes the hydrogen bonding employed in protein folding. The characteristics of how a particular protein arranges or folds itself play a critical role in determining its biochemical properties. These effects are not roadblocks on the highway of life—more like sharp bends that require additional chemical attention to negotiate.

Some scientists have a problem with the possibility of life on Titan because the liquid medium is likely a mixture of methane and ethane—both of which are nonpolar molecules. Many of the polar salts, proteins, and organic chemicals that play a role in Earth life would not dissolve in methane and ethane. Instead, they would precipitate out of the liquid and collect as sludge at the bottom of the pond. But there are just as many nonpolar organic chemicals that do dissolve in liquids such as methane and ethane. Just ask physical organic chemists in which liquid they perform most of their lab experiments—the answer is that much of the time it is not water. Those chemicals that dissolve well in liquid methane and ethane also have greater freedom to exploit weak hydrogen bonds—a bond that may be particularly well suited to low-temperature environments.

Could this alternative, nonpolar organic chemistry play a role in alien life? The answer is clearly yes. The more important question is how—what are the critical biomolecules we should be on the lookout for? To answer that question, we need to go to Titan and analyze several flasks full of liquid from some "cold little pond" of methane—a Titanian echo of Darwin's warm little pond on the ancient Earth. As we shall see, though, that will take a lot of bucks (billions and billions, actually) and several decades to achieve. In the meantime, what aspects of Titan can we synthesize on Earth to gain a head start on understanding the chemical—and just possibly biochemical—environment that we may encounter?

Goldilocks on Titan

Might Titan be "just right" for life? One way to approach this question is to take the same concepts that have been used to understand the emergence of life on Earth and apply them instead to Titan. Perhaps the most effective laboratory approach used to understand the chemical landscape of the early Earth is the Miller-Urey experiment.

One reason why Titan fascinates astrobiologists is that the present-day atmospheric composition is not so different from what we believe existed on the early, nonliving Earth. The important components are the presence of hydrogen (which likes to share electrons) and the relative absence of oxygen (which is electron greedy). In the atmosphere of both Titan and the early Earth, hydrogen is the great facilitator, donating electrons for the synthesis of new organic chemicals. Oxygen is the great reactor—always seeking to bind to new elements and molecules to sate its electronic appetite.[7]

So how might one simulate Titan in a lab? Achieving the right chemical mix is not too difficult—95 percent nitrogen, a small percent of methane, and a sprinkling of carbon monoxide. But what about the energy to power the chemical reactions? The surface of Titan is darkened by thick layers of atmospheric haze—no ionizing radiation there. In addition, we have no evidence for lightning on Titan—no vital spark to energize chemical reactions. But the upper atmosphere of Titan—above the haze—is a rich photochemical environment. We know this because we have observed with *Voyager* and *Cassini* the many chemical by-products—at least the simpler ones—resulting from the photodissociation of methane by solar photons. The challenge is therefore to simulate this high-altitude environment—combining extremely low pressure and ionizing radiation—while ensuring that the resulting chemistry occurs in the gas phase and not on the walls of the pressure chamber.[8]

This Titan-esque version of the Miller-Urey experiment differs from Stanley Miller's original experiment in one critical aspect—no liquid. Miller—and those who emulated him—circulated organic chemicals through a flask of liquid water that represented Earth's oceans. In our version of the Titanian atmosphere, all the chemistry takes place in the nitrogen-rich gas phase. So given this difference—which may prove critical—what have been the results from this new generation of "alien" Miller-Urey experiments? Perhaps surprisingly, changing the recipe and conditions to reflect Titan rather than ancient Earth does not change the flavor of the results. The Miller-Urey experiment concocted according to Titan's recipe produces many of the same amino acids and nucleobases (C, A, G, T) present in living creatures on today's Earth. The main difference is the amount of stuff produced—Miller could scrape it from the inside of his flask and use relatively simple instruments to identify the principal molecular species. The amount of organic residue produced in the Titan gas-phase experiment is generally much, much less and requires a sensitive analysis to sniff out traces of amino acids and nucleobases.

You may well be wondering how organic molecules synthesized in the upper atmosphere of Titan could make it down to the surface to join the imagined prebiotic house party. Truth be told, we just don't know. The most interesting organic products of this experiment—in this case, the amino acids and nucleobases—are relatively much heavier than the molecules making up the bulk of the atmosphere. At present, we assume that they would diffuse down to the surface—a physically reasonable view, yet one that lies at the mercy of our almost complete ignorance of the dynamics of Titan's atmosphere.

But what does this mean for life on Titan? The production of the same amino acids and nucleobases in labs simulating both Titan and ancient Earth may well be telling us that these molecules are common waypoints on the winding reaction pathways followed by organic chemicals in each environment. Although some of these amino acids and nucleobases have been incorporated into Earth life, it may well be highly unlikely that a similar route is being followed on Titan—mainly since these particular organic chemicals do not dissolve well in liquid methane or liquid ethane.

We may have overlooked the critical molecules produced in our Miller-Urey experiment simply because we didn't know any better—our view of life remains biased toward Earth life and its constituent molecules. Regardless, we can say that Titan is a rich, chemically active environment with striking similarities to the conditions of the early Earth. The prevailing scientific view is that Earth life emerged naturally from just such an environment. In this sense, Titan is "just right" for life. But remember that we have little idea how Earth went from being "just right" for life to actually giving rise to it. Given all that we are discovering about Titan's rich chemical landscape, it is clearly a potential habitat for life that is difficult to dismiss.

When Is Cold Too Cold for Life?

Is Titan just too cold for life? Can we rule out the possibility of life on the grounds of low temperature? Here again the answer is no! The bottom line is that life is not powered by temperature; life is powered by chemical reactions.

It is true that chemical reactions proceed at a faster rate at increased temperature. What we sense as temperature is simply the random motion of the particles (atoms or molecules) that make up the medium. Higher temperature means more motion (velocity), which means a proportionally shorter amount of time for one atom or molecule of interest to encounter another and react with it. There is no temperature cutoff for this statement. If you have a sample at 300 Kelvin, it has twice as much thermal energy as the same sample at 150 Kelvin.

The water from which you and I are mostly made would freeze rock hard well before we reached this limit, so you could say that below a certain low temperature, conditions would be too cold for Earth life. True enough, but that is just because Earth life is water based. Choose another liquid medium—say, ammonia, methane, or ethane—and a new temperature frontier opens up for life. You would have to come up with the exact organic chemistry for this new kind of life, but the basic fact is that much of known organic chemistry would still be available to you. You can even push the limits of Earth life by dosing antifreeze with a small amount of water and seeing whether the enzyme chemistry used by all Earth life still proceeds. It does—down to a temperature of −100° Celsius (173 Kelvin). So while low temperatures will result in longer timescales for potential biochemical reactions, they will not hamper them in any fundamental way.

If we drop our Earth-centric view that temperatures around 0°–100° Celsius are "best" for life, then we may realize that much lower temperatures can be positively beneficial for life to proceed. In particular, Titan's organic chemistry may make use of hydrogen bonds—which are weaker than covalent bonds—to form a wider range of stable chemical relationships than would be available at higher temperatures. Low-temperature Titanian life may march to the beat of a slower drummer than heat-loving Earth life, but as long as conditions on Titan are stable, these slower rhythms will not impede either the prebiotic path to life or the development of life itself.

An Inconvenient Truth

The biggest challenge to discovering life on Titan may well be figuring out how to recognize it. Even on Earth, scientists across a range of disciplines are unable to agree on a single, fundamental definition of life.

Rather than consider the different definitions employed by zoologists, botanists, chemists, and molecular biochemists, among others, let's focus on the astrobiologists. Think about some of the previous occasions when astrobiologists have performed direct searches for life, for example, within the context of the *Viking* missions or the analysis of ALH84001. Each group of scientists was using a particular definition of life, and they constructed their experiments or performed their analyses to test exactly that hypothesis.

Let's start with *Viking*. In each of the *Viking* biology experiments, nutrients were added to Martian soil samples under a variety of conditions. During the mission planning, scientists decided that the presence of gases given off during the experiment was to be accepted as a good indicator of metabolic activity by Martian microbes. The definition of life here was "I metabolize, therefore I am." What about ALH84001? The scientific team searched for microscopic—even nanoscopic—physical structures that could be identified as fossil cells. In this case, the definition of life was "I organize, therefore I am."

So if we go to Titan—for the moment, let's assume it will be robotically—what should be our working definition of life? We have considered the merits of defining life as a self-sustaining chemical system capable of Darwinian evolution. But given this definition, what would we look for and what tests would we perform?

One of the main proponents of this evolution-led search for life is Steven Benner from the Foundation for Applied Molecular Evolution. His idea is that life on Earth uses three critical biomolecules: DNA, the repository of genetic information; RNA, the combined messenger and builder; and proteins, the workers. Perhaps more important, he has asserted that the atomic structures of these terrestrial biomolecules follow some simple principles that may be universal in nature. Identifying alien molecules constructed on similar principles could provide a route to detecting life.

Among the simplest principles to grasp is the polyelectrolyte theory of the gene. It holds that the critical feature making DNA a good molecule for encoding information is the repeating negative charge present on the phosphate groups that make up the backbone of all DNA molecules. Since like charges repel each other, this tends to stretch out DNA into long strands—a much better arrangement for sequential chemical interpretation by RNA than a messy, folded molecule. In addition, the repeating backbone charge dominates the large-scale chemical properties of the DNA molecule, for example, the fact that it dissolves well in water. Changing the sequence of the C, A, G, and T nucleobases, which spell out the words of life, does not change the large-scale chemistry of the molecule—a feature that, it could be argued, is a critical requirement for maintaining freedom of genetic expression.

So the challenge here is to construct experiments, inspired by the laboratories of molecular biologists, that would be simple and robust enough to journey to Titan and search for these characteristic molecular structures in situ. This is a neat new direction—a molecular biologist's definition of life. The practical astrobiologists among you probably recognize that some combination of the above three ideas—metabolism, cellular structure, and molecular structure—would be required to construct a comprehensive Titan biology package.

Who Ate All the Acetylene?

Titan constantly challenges our ideas of atmospheric and lunar chemistry. Some ideas we think are pretty good: the abundant atmospheric methane is converted into more complex organic chemicals via reactions with sunlight. Other aspects of Titan's atmosphere reveal the limits of our knowledge: if all that methane is being converted into more complex chemicals, why doesn't it get used up?

At the rates of photochemical reactions involving methane, the methane currently present in Titan's atmosphere will be consumed entirely within about fifty million years or so. That's the blink of an eye in solar system timescales. So what is replenishing the methane? Cryo-volcanism, which predicts the venting of methane gas from subsurface reserves, offers one explanation—and a very sensible one, too, although clear examples of volcano-like structures on the surface of Titan have yet to be discovered. Another source of methane may well be biological—and it is a sufficiently interesting story to bear telling.

In 2005, the same Chris McKay who taunted us with Rare Titan published a short paper that recognized the presence of a lot of potential metabolic fuel in the atmosphere of Titan. One of the simplest—and most energy-rich—diets available to Titan residents would be to combine acetylene with hydrogen to produce energy and two methane molecules, that is, $C_2H_2 + 3H_2 = energy + 2CH_4$. What do you note from this reaction? Acetylene and hydrogen are consumed by surface-dwelling Titan life, and methane is produced.

Okay—but how does this compare to what we measure in Titan's atmosphere?

Perhaps surprisingly, atmospheric observations from *Voyager* and *Cassini* seem to be consistent with the effects of this simple acetylene metabolism. Two studies have shed light on the curious chemistry on Titan. The first study uses data from the *Voyager* and *Cassini* missions to measure the relative amount of molecular hydrogen at high altitudes—where it is believed to be created via the interaction of methane and sunlight—and close to the surface. The scientists performing this study were surprised to detect more hydrogen at the bottom of Titan's atmosphere than at the top—not just because hydrogen is created at high altitude, but also because it is the lightest atmospheric gas. This measurement implies a large downward flow of hydrogen from the high-altitude haze layers—about as much hydrogen, in fact, as escapes into space. So where does it go? Something must be using up the hydrogen as it flows to the surface, but whether this is hydrogen-loving life or a more prosaic (yet still interesting) surface chemical reaction, we just don't know. We do know that the surface dust of accumulated organic material is devoid of acetylene. One of the most common by-products of the photo-destruction of atmospheric methane, acetylene should steadily rain down on the surface of Titan. Yet *Cassini* observations have failed to detect its chemical fingerprint in the spectrum of light reflected from the surface.

So is this unambiguous evidence for acetylene-munching microbes basking on the smog-shrouded shores of a Titanian lake? Certainly not—a good astrobiologist should use life as the conclusion of last resort when confronted with new evidence. A common process might well link the fate of downward-flowing molecular hydrogen and the absence of acetylene. It could be that some nonbiological surface reaction brings acetylene and hydrogen together and synthesizes new organics—and it is interesting in this respect that much of the surface of Titan is coated in an organic tar of unknown chemical composition.

Yet the clues remain compelling. Basic life will not wave to our passing spacecraft and peering telescopes. Instead, we will be offered mysteries, anomalies, things that don't add up. The possible presence of methane on Mars, the fate of acetylene and hydrogen near the surface of Titan—these are some of the clues that we have uncovered, and we ignore them at our peril. What we need in this case is a more detailed understanding of the chemistry of Titan—a need that will ultimately require us to visit.

Mission to Titan: To Boldly Float

Given all that we have learned of the environment on Titan—and more important, of what we remain ignorant—what science should a future Titan mission attempt? What will the spacecraft look like? What technology will it require to achieve our goals? There are a number of mission plans currently under development—though, unfortunately, all are currently languishing in the limbo of missions that are good but not good enough to make it to funded reality.

Can we look beyond the valley of NASA's and the ESA's current budget woes to the sunlit uplands of a future Titan mission? Of course we can—and this is the task of those who advocate Titan as the destination of the next ambitious, flagship mission to the outer solar system. Many of the core ideas of a future Titan mission have been brought together under the umbrella of the Titan Saturn System Mission (TSSM)—to enter the Saturnian system and execute multiple flybys of Titan and Enceladus en route to a stable orbit about Titan. From its high orbital perch, the spacecraft would then dispatch two craft into Titan's atmosphere.

The first would carry a six-hundred-kilogram science package and would deploy underneath the canopy of a hot-air balloon— romantically referred to as a *montgolfière* in the NASA-ESA mission overview. This is exceptionally ambitious, but it also makes a lot of sense: the atmosphere of Titan is dense and cold. The excess heat produced by the onboard radioisotope thermoelectric generator would produce a warm, buoyant bubble of air. The nominal lifetime of the *montgolfière* would be six months, during which time the circumlunar winds would carry the ten-meter-diameter balloon once around the moon at an altitude of ten kilometers.[9] The main science goals for the balloon-borne probe would be to image the surface of Titan at wavelengths and spatial resolutions inaccessible to orbital craft—for example, wide-angle visual imaging down to one-meter resolution. The probe would use a mass spectrometer to sample the atmosphere in situ, to determine its chemical components and how they vary with location. It would certainly mark the greatest balloon journey ever imagined.

The second craft would be deployed to one of Titan's great lakes, to float on the lake surface. This probe would be a chemist par excellence, with an onboard mass spectrometer capable of measuring the properties of molecules containing up to ten thousand atoms. A lamp and a visual imaging camera would, respectively, illuminate and capture the scene on the lake surface. The clock would be ticking on this phase of the mission. The current design calls for chemical batteries rather than a costly and rare plutonium power source. With a nominal battery life of nine hours, of which six would be used up in the descent phase, only three hours would be available to power the craft during its analysis of the lake waters.

Above all, the Titan Saturn System Mission is a chemical explorer—it will characterize the chemistry of the Titanian atmosphere and lakes and (I am sure) astonish us. Will it reveal life? Only if we are confident of confirming life from the presence of molecular order. Though it may well reveal the raw materials which life—if present—will have to work with.

If, in the spirit of this book, you were planning a Titan exploration mission, would you be satisfied with these goals—or would you try to push further? If you were truly ambitious, you could imagine elements of a Titan biology package similar to that deployed on *Viking*. You could try to anticipate the salient features of a Titanian metabolism and construct an appropriate chemical test. But the lesson learned from *Viking* is that until you have a broader understanding of the chemical environment in which you are performing such experiments, your results will be at best ambiguous regarding the presence of life. Because of budgetary restrictions, the one biology experiment that did not fly on *Viking* was a "Wolf trap," named after its creator, Wolf Vishniac. His idea was to add Martian soil to a vial of water and to measure any changes in turbidity (cloudiness) due to microbial growth. Such Wolf traps are a common experimental tool used in the biological exploration of the Dry Valleys of Antarctica. It is a neat and simple idea, "I grow, therefore I am," and one that could be adapted to test for microbial growth in the lakes of Titan— though one would add the sample to a vial of liquid methane rather than water.

One could also imagine that the old adage "seeing is believing" would present a powerful argument for including a microscopic imager to determine exactly what potential biological structures the lake waters might contain. We like to imagine the moment of scientific revelation when a child on Earth looks for the first time through a microscope into a drop of lake water and discovers the realm of tiny creatures within. But can you imagine the sense of anticipation if we included a powerful microscope on a future Titan lander? Getting samples ready to be viewed might well prove tricky, but the resulting images and time-lapse movies would provide a powerful test for the presence of cellular life.

Given these limitations and ambiguities, why am I not telling you to get all gung ho and fire off a Titan sample return mission—isn't that really what we need? Yes, sure, but you will need an awful lot of dollars to do it. Attempting a sample return mission to Titan makes a Mars mission look easy. You have to work with what you've got and, if that is $4 billion or so, that means a mission like the TSSM—which currently clocks in with a price tag of $2.5 billion.

Do Not Go Gentle into That Good Night

When *Voyager 1* executed its flyby of Titan on November 12, 1979, it gave up its chance to encounter additional planets in the outer solar system. With the prospect of a close-up look at Titan, a not-to-be-missed opportunity, mission controllers directed *Voyager 1* into a close flyby that swung the tiny probe out of the plane of the planetary orbits and out of the solar system. Hot on its heels, *Voyager 2* sailed past Saturn on August 25, 1981, yet maintained a course through the outer solar system that brought later, astonishing encounters with both Uranus and Neptune.

What fate awaits these voyagers? Each *Voyager* probe is currently passing through a turbulent region of the solar system dubbed the heliopause—the region of space over which the particle pressure of the solar wind drops dramatically to the lower pressures of interstellar space. Some commentators have called this region "the edge of the solar system." In fact, it is only the first of several waypoints on our journey.

Voyager 1 is traveling at a velocity of fifty-six thousand kilometers per hour, or three astronomical units[10] per year. It will pass beyond the heliopause and, perhaps a thousand years from now, will traverse the icy realm of the Oort cloud, an unknown and unseen realm of ancient comets. About halfway to the nearest stars, the *Voyager* probes will pass beyond the gravitational influence of the sun and join the great procession of stars in orbit about our galaxy. It will ultimately take some ninety thousand years for each spacecraft to cover the distance to the nearest star system, Alpha Centauri—4.3 light-years from the sun.[11]

We have taken our first steps into a wider world. Will new technology allow us to catch up and pass *Voyager 1* en route to the nearest stars? Will we even exist as a species ninety thousand years from now, having matured beyond our troubled youth? Deep questions indeed. What we can say is that even the nearest stars are almost unimaginably distant in relation to our current space-faring ability. Nonetheless, it is to them that we must now turn, using telescope observations rather than spacecraft, to explore their planetary systems and prospects for life.

Do you remember, way back in chapter 1, when I encouraged you to step outside and look at the stars in the night sky? Each of the several thousand stars visible to the naked eye is a sun much like our own. Beyond them, fainter and more distant, are the stars, up to four hundred billion of them, that make up our Milky Way galaxy. But if our sun is typical, one among the billions of stars in the Milky Way, what of the planets that make up our solar system? Are they typical? Does each star we see in the night sky host its own retinue of planets?

This chapter describes the discovery of exoplanets, planets in orbit around distant stars. Are they strange new worlds akin to those dreamed of in science fiction, or familiar worlds resembling those of our own solar neighborhood? It turns out they are both. Despite the fact that most exoplanet detections are indirect—the light from the planet is lost in the glare from the much brighter parent star—we are able to measure some of their fundamental physical properties. Are they gas giants like Jupiter or terrestrial worlds similar to Earth? Would they be warm and welcoming, or would we encounter hostile extremes of temperature?

The story of exoplanets is one of pure discovery: thousands of new worlds whose existence was unknown to us even twenty years ago. Our ever-inquisitive inner astrobiologist poses yet more penetrating questions: How can we assess their potential as habitats? Might these planets host life? How could we confirm its presence? A great scientific journey lies before us. But there is clearly one catch: the stars, and their planets, are very far away. When we consider the search for life on exoplanets, we realize that our methods must change. We are fundamentally removed from distant planets and any life they might host. Dispatching robotic probes to attempt audacious sample return missions is no longer a practical option. Instead, we must use telescopes and their detectors to perform remote observations.

To Catch an Orbiting Planet

So how do we detect exoplanets? In chapter 1, I introduced the first exoplanet discovered, 51 Peg b, and described how it was detected through its gravitational influence on its parent star. The orbital motion of the planet causes the star to execute a small orbit of its own, motion that can be detected using a sensitive spectrograph. This technique is known as the stellar radial velocity method, or the Doppler wobble technique, and since 1995 it has been used to discover several hundred new exoplanets.

In this chapter I want to tell you a slightly different story, focusing on a different method for detecting exoplanets. This method, known as the planetary transit technique, is neither better nor worse at discovering planets than the Doppler wobble technique (or any of the other techniques that page limits require me to skip). Yet the story of the planetary transit technique, its development and successes, is a good one, and I can never resist a good story.

Twinkle, Twinkle

The stars we observe in the night sky appear to twinkle from the effects of turbulence in Earth's atmosphere. If you were to shift your perspective and observe the stars from space—as we do with a host of orbiting telescopes—the shifting, restless images would become still.

Yet some stars lead intrinsically variable lives—pulsations in their vast atmospheres of plasma cause their brightness to vary with time like some slow stellar heartbeat. But even when we focus our attention solely on well-behaved, apparently unvarying stars, we sometimes detect faint, almost imperceptible flickerings—rhythmic changes in their brightness as regular as the ticking of a clock. This is the signature of a planet. The rhythmic change is a small dip in the brightness of the parent star caused as the planet passes in front of it. We call this event a transit. In practice, it is very much like the solar eclipses we view from Earth—though in this case it is a planet rather than our moon that is blocking the light from the star.

Planetary transits occur in our solar system. As viewed from Earth, Venus regularly passes across the disk of the sun, taking approximately seven hours to complete its passage. For the duration of the transit, Venus blocks a small amount of sunlight from reaching Earth. How much exactly? As viewed from Earth, Venus looks like a circular black disk as it passes in front of the sun. Basic geometry gives the area of a circle as pi times the square of the radius—in this case, the radius of Venus, or about 6,000 kilometers. The area of the disk of the sun is equal to pi times the square of the radius of the sun—which is 700,000 kilometers. The fraction of sunlight blocked during a transit of Venus is just the square of the ratio of these radii—in this case, approximately 100 squared, or 1 part in 10,000.

So much for Venus as viewed from Earth. But what would a distant observer of a larger planet, such as Jupiter, detect if it passed in front of our sun? Jupiter is about ten times as large as Venus. The fraction of the sun's disk blocked by a Jupiter transit would be 1 part in 100, or 1 percent. It turns out that this brightness change is just about detectable using current ground-based telescopes and their detectors. Thus, it should come as no surprise that the first exoplanets detected by the transit technique were Jupiter-sized worlds orbiting sun-like stars. The sensitivity of transit planet searches has increased dramatically, and as we shall discover, present-day space-based surveys now routinely detect Earth-sized worlds orbiting sun-like stars.

But what if the orbit of a planet was such that the planet did not appear to pass in front of its parent star? What if, from the perspective resulting from our particular viewing angle, we looked down on the planet running rings around its star? From our direction in space, the planet would never pass in front of the star—we would never see a transit in this system.[1]

Quite right. We observe transits as a result of chance alignments, where the planetary orbits are inclined so that—from our point of view—they just happen to pass in front of the star. Does that matter in any fundamental way? Not one bit. Depending on the size of the star, the size of the planet, and its orbital distance, we might expect 10 percent of distant planets to transit in front of their parent stars.

These 10 percent of planetary systems will not be special or different in any other way. It is purely chance that governs whether we can observe their transits or not.

What can we learn about a planet from the repeating eclipses of its parent star? First, we can determine the planet's orbital period. If the transits are observed to occur every twenty days, then it must take the planet twenty days to complete one orbit about the star. It really is that simple. By what fraction is the starlight dimmed? This tells us the projected area of the planet compared with the projected area of the star—which in turn tells us what their relative radii are. The important point to note is that stars obey a very clear set of physical laws. Once you know a star's luminosity and the temperature of the gas in its outer layers, you can calculate a very accurate radius. You can then use the properties of each transit to work out the radius of the planet. If you can measure the planet's mass via the Doppler wobble, then you can calculate its density as well. This offers a fundamental clue to the physical nature of the planet—dense and rocky or tenuous and gassy?

So, relatively simple observations of planetary transits can tell you the planet's orbital period and its radius (plus mass and density). The same observations used to detect the transit can also tell you the kind of star it orbits—more massive, luminous, and hotter than our sun, or smaller, fainter, and cooler? We can in fact learn a great deal more about the planet and its orbit than that. But to go further I will have to introduce you to my friend Kepler.

Kepler the Man

Johannes Kepler lived in what today we call Germany and Austria, from 1571 to 1630. He was an astronomer and mathematician, a contemporary and peer of Galileo. Like Galileo, his life was profoundly affected by the religious turmoil and war that engulfed Europe during this period. I want to explain why Johannes Kepler is a hero among scientists—even if he does not share the wide repute of Galileo. Kepler is a hero—at least to me—because he was perhaps the first scientist who realized that his ideas about the universe were wrong. Wrong because they did not agree with observations. His path to greatness is notable not so much for what he discovered as for what he gave up.

To say that Kepler was fascinated by the motions of the planets is a considerable understatement. He was obsessed. Kepler's first major theory describing the absolute motion of the planets in space used circular orbits whose radii were based on a set of nested "perfect," or Platonic, solids.[2] This model was elegant and brought a pleasing mathematical harmony to the newly sun-centered solar system. It was also wrong. Kepler realized his error by following the method employed by scientists ever since—he used his model to predict the apparent position of the planets in the sky and compared the predictions to the best available data. It is worth emphasizing that the work required to achieve this was computationally tedious in a manner that can barely be imagined in today's processor-powered world.

Although Kepler was never much of an observer himself, he became (following a bitter dispute) the custodian of the great catalogue of astronomical measurements painstakingly compiled by his predecessor and mentor, Tycho Brahe. Kepler worked for Brahe up until the latter's death in 1601, and these two very opposite characters shared no liking for each other. Yet it is important to emphasize that Kepler was astute enough to trust Tycho's work. The typical error in the position of an astronomical object in Tycho's catalogue was around two minutes of arc.

Given the mathematical labor involved in computing planetary positions, Kepler focused his efforts on Mars: as viewed from earth, Mars regularly executes a retrograde or reverse loop in its path across the sky. It seemed to offer the keenest observational challenge to any successful theory—and Kepler's carefully computed circular orbit failed this challenge. His predicted position of Mars differed from the observed position by eight minutes of arc, or approximately one-quarter of the diameter of the full moon. Though disappointed that his years of labor had come to naught, Kepler realized that observations counted. Kepler could have ignored that discrepancy or rationalized it away. He could have questioned the accuracy of the data. He could have decided that a really elegant theory should be conclusively disproved before being discarded. But Kepler firmly took his place as perhaps the first modern scientist when he realized that details matter—and his theory got those details wrong.

Kepler returned to his desk, disappointed yet undeterred. His long computational quest had returned him to his starting point: the troublesome orbit of Mars. Kepler had already brushed against elements of the true answer—he just hadn't realized it. Could the orbit be elliptical in shape? Kepler was aware that both ellipses and circles formed a continuous family of possible orbital trajectories—each is a conic section, formed by a two-dimensional slice though a three-dimensional cone.

Returning to the idea of elliptical orbits, Kepler unexpectedly found that the motion of Mars was a perfect fit. The other known planets took their own places in the solar system, each moving along its own elliptical orbit. He presented this discovery as his first law of planetary motion: planetary orbits are elliptical, with the sun at one focus. Each planet moved according to a similar rhythm—faster when closer to the sun and slower when more distant. This became his second law. Ultimately, Kepler expressed the mathematical harmony of planetary orbits via his powerful third law: the square of the orbital period is proportional to the cube of the distance from the sun, or $P^2 \propto a^3$.

Kepler died in 1630; his laws were resurrected some fifty-seven years later when, in 1687, Isaac Newton published his *Mathematical Principles of Natural Philosophy*. In this book, Newton put forward his theory of universal gravitation in one of the most astonishing acts of creative thought ever accomplished. Kepler's laws fascinated Newton. The predicted positions of the planets remained as accurate as the day the laws were formulated. Yet why? What unseen power moved the planets about the sun? In answering this question, Newton created his theory of universal gravitation, whereby the unseen force of gravity results from the product of the masses of the two bodies divided by the square of their distances. Newton demonstrated that Kepler's laws should hold not just for the solar system but also for any system of planets in orbit around any star. In simple terms, you just need to know the mass of the parent star—it alone sets the scale of the planetary orbits.[3] Together, Newton and Kepler traced the motion of the planets about not just our sun but all stars that host planets.

We can now at last understand the next piece of the puzzle extracted from planetary transit observations: once you know the orbital period of the planet and the mass of the parent star, you can use Kepler's third law to compute the orbital radius. You have measured the scale of a new planetary system.

Kepler *the Spacecraft*

The observed properties of planetary transits provide a keyhole through which to view distant solar systems. Yet astounding precision is required to achieve this. The transit of an Earth-sized planet in front of a sun-like star causes a brightness change of 1 part in 10,000, or 0.01 percent. To be really sure that you are observing a transit and not just a noise blip, you need to perform measurements with an uncertainty about one-fifth as large as the expected signal, or 0.002 percent.[4] That's a challenge when even the best ground-based measurements of stellar brightness today rarely achieve a precision better than 1 percent.

Transits need precision, and that means going into space. Observing beyond Earth provides a number of important advantages. You have left Earth's atmosphere, its turbulence and latent glow, behind. You can observe continuously, without regard for Earth's diurnal rhythm. And finally, stable viewing conditions, combined with low-noise digital detectors, allow you to perform photometry[5] of the highest precision.

In 1984, a small team of researchers led by William Borucki of the NASA–Ames Research Center in California was the first to put all these ideas together. They conceived of a space observatory that would observe a single patch of sky, containing up to 160,000 bright stars, continuously over a four-year period. The probability of chance

EXOPLANETS

alignments meant that even if every star hosted a planet, one might expect at most 10 percent, or 16,000 stars, to show transits. Observations would be performed with enough precision to detect the transit of an Earth-sized planet in front of a sun-like star—and if this distant Earth took one year to orbit its parent star, then the four-year mission to discover strange new worlds would detect up to four transits in this one system—just to be sure of its nature.

It ultimately took over twenty years for *Kepler* the space telescope to fly and achieve this promise of discovering new worlds. NASA rebuffed mission proposals numerous times, and it is a credit to the dedicated team of scientists that each technical query and doubt was overcome with successive laboratory and on-sky tests.[6] *Kepler* was selected as a NASA Discovery-class mission in December 2001. The cost? Six hundred million dollars—which buys you the hardware, launch, and science analysis back on Earth. The science returns to date make that investment look pretty impressive.

The *Kepler* space mission was launched on March 6, 2009. The nominal mission lifetime was 3.5 years, but with the possibility to observe for a maximum of six years. *Kepler* is not a huge spacecraft by any measure; its primary mirror has a diameter of 0.95 meters. *Kepler*'s real gem is its detector, a ninety-five-megapixel camera of superlative electronic performance. The telescope points to a star field in the constellations of Cygnus and Lyra, and its view of the sky covers 105 square degrees. That is an impressive chunk of sky—the moon has an angular diameter of half a degree, so *Kepler*'s field of view is equivalent to a square big enough to fit twenty-one moons along each side.

Why so big? Because this field contains 160,000 stars—bright enough to yield the required photometric precision, yet not overlapping with one another or projected onto background galaxies. *Kepler* observes this star field every six seconds before storing and processing each image. Not all the raw data is kept, since *Kepler* would soon run out of onboard disk space. Furthermore, the rate at which data accumulates greatly outstrips the rate at which it can be downlinked to Earth. Instead, *Kepler* measures the brightness of each of the 160,000 target stars and stores only that information. Batches of brightness measures from each six-second image are then merged into a single three-minute average. *Kepler* stores this greatly compressed information, and only once a month does it send a digital postcard back to Earth with the brightness data for every star. These are the raw data gleefully received each month by the *Kepler* mission scientists.

Kepler followed this routine for just over four years, steadily and comprehensively measuring the brightness of 160,000 stars, once every six seconds. It has been an exercise in patience and discipline of which Kepler the man would be rightly proud. Unfortunately *Kepler*'s discovery mission came to an abrupt end on May 11, 2013, when the second of four reaction wheels aboard the spacecraft broke. Each reaction wheel functions as a fine-tuned stabilizer that allows the telescope to remain aligned along a particular axis. Since space has three dimensions (or axes), a minimum of three functioning reaction wheels are required to point the telescope at a particular location in the sky. Furthermore, since the accuracy of the telescope pointing is a critical component that ensures the overall precision with which *Kepler* can measure the brightness of stars, the failure of two wheels essentially left *Kepler* rudderless and defunct.

The main planet-discovery mission has ended, but the demise of *Kepler* itself is greatly overstated—the spacecraft is still able to monitor new areas of the sky by the ingenious use of the radiation pressure of sunlight for stabilization. There will certainly be new planet-finding missions in the future, yet it is no exaggeration to say that *Kepler* has provided us with a leap in our knowledge as great as any space mission past, present, and possibly future. So what have we learned? What do these brave new worlds look like?

Hot Jupiters . . .

The discovery of exoplanets has required us to create new adjectives to describe them—new words for new planetary environments. Hot Jupiters and super Earths now accompany the stars of our Milky

Way. The most striking fact that has emerged from the exoplanets discovered to date—from all techniques, not just from *Kepler*—is how varied they are. As mentioned earlier, 51 Peg b was the first hot Jupiter to be discovered; now I want to introduce you to Upsilon Andromeda b, or "Ups And b" for short. The star Upsilon Andromeda is located some forty-four light-years from Earth. It is an F-type star, somewhat hotter and more luminous than our sun, and as luck would have it, is easily visible to the naked eye.[7] Ups And b was discovered in 1996, one year after 51 Peg b, and it is another example of a hot Jupiter—this puzzling new type of planet. But what measurements justify this description? In particular, how do we determine the temperature of a planet?

When we consider the basic radial velocity data for the Ups And system, we can calculate that the mass of Ups And b is about one-half that of Jupiter and that it takes 4.6 days to orbit its parent star. If the planet's mass is similar to that of Jupiter, then we can assume it is a Jupiter-like world—but why is it hot? To understand why, we have to think back to Kepler's third law and the relationship between orbital period and orbital radius. If the parent star has a mass similar to the sun's, then we can think of the orbit of Ups And b in terms of the scale of our solar system. So with an orbital period of 4.6 days, the orbital radius of Ups And b must be about one-twentieth (more exactly, 0.06) of an astronomical unit (AU). That is about one-eighth as far from its star as Mercury's orbital radius. It is clearly going to be one hot planet, but how hot exactly?

To calculate the surface temperature of a planet, we assume one very important thing: that it balances the energy received from its parent star by that reradiated into space. Depending on the temperature of the parent star, the fraction of starlight absorbed by the planet (as opposed to that reflected) and the orbital distance of the planet, one can calculate at what surface temperature this energy balance occurs. For Ups And b, the equilibrium temperature is over 1,400 Kelvin, which is one hot Jupiter.

One of the wonderful surprises contained within the Ups And system is that when you subtract the effect of planet b from the

Doppler wobble of the parent star, the star still shows a detectable radial velocity variation—there must be more planets. It turns out that Ups And contains four Jupiter-mass planets, all orbiting within the scale of Jupiter's orbit in our solar system. The other three are more distant than planet b and consequently display successively lower equilibrium temperatures. I hope that at this point a light has just switched on and that you can see the link between equilibrium temperature and the idea of the habitable zone. Although the habitable zone can be defined in a number of ways—by using first-year university physics alone, all the way to running more complex models assuming a particular atmospheric composition—perhaps the simplest definition is that it is the orbital radius about the parent star at which the planetary equilibrium temperature is between 100° and 0° Celsius.

... and Super Earths!

One of the biggest surprises in the menagerie of planets discovered by *Kepler* was the large number of worlds discovered with radii between one and four times that of Earth. In our solar system, both Uranus and Neptune have radii about four times that of Earth. Their masses are fourteen and seventeen Earth masses, respectively. There exists nothing else in our solar system between their size and that of Earth—yet in the *Kepler* detection catalogue, these intermediate-size planets are among the most common. What would these planets look like? Would they be terrestrial worlds more massive than Earth—so-called super Earths—or would they be scaled-down Jovian worlds, wimpy Neptunes?

One of the most extreme examples of a super Earth is a planet called Kepler 10c. Discovered by the *Kepler* probe as a transiting planet, its radius is 2.3 times that of Earth's. The radial velocity of the parent star reveals the planet's mass to be seventeen times as large as Earth's—almost exactly the same as Neptune. Knowing the mass and the radius, we can calculate the density of Kepler 10c. It is 1.3 times as dense as Earth and is therefore no gas giant; it must be a super-super Earth. But what determines whether a seventeen-Earth-mass planet is rocky (like Kepler 10c) or gassy (like Neptune)? We just don't know—at least not yet.

In hindsight, it would have been a little disappointing, boring even, if we had discovered that our solar system was typical—only one of many similar systems constructed according to a repetitive template of planetary rules. Though bewildering, our current view of planetary science—as revealed by exoplanets—bears more resemblance to a loosely controlled riot of planet-making possibilities. And it is much the more interesting for it!

A Plurality of Worlds

We don't have an official collective noun for planets.[8] A "system of planets" is functional but not particularly inspiring. A "plurality of planets" has a nice ring to it. The phrase is often attributed to Giordano Bruno, a sixteenth-century priest and philosopher who speculated that the stars were suns and that each was accompanied not just by planets but by inhabitants as well. Cosmic pluralism was a widely embraced idea through the scientifically enlightened centuries that followed—only to enter apparent (and thankfully short-lived) disrepute in the aftermath of Lowell's popularization of life on Mars. The results from the *Kepler* data have revealed a true cosmic pluralism: planets are common. So common in fact that I almost opted for different collective noun—a bewilderment of planets!

When *Kepler* commenced its mission, we already knew of the existence of some 332 exoplanets. Astronomers had detected these planets by a variety of techniques, not just transits. At the end of 2014, we knew of 1,849 planets, about half of which (923) were discovered by *Kepler*. Astronomers consider these planets "confirmed," which means that they have measured each planet's mass either by stellar radial velocity observations or by subtler techniques such as transit timing variations.[9] Beyond these 923 confirmed planets discovered by *Kepler*, there are more than 2,500 candidate planets. Based on the success rate with which such candidate planets have previously been confirmed, more than 90 percent of these are likely to be real.

Let me pose a fundamental question. How many planets orbit a typical star? Has *Kepler*'s census revealed an answer? As of 2014, *Kepler* has discovered 3,533 candidate planets in orbit about 2,658 stars. Approximately one in five stars in this sample hosts more than one planet. The analysis of *Kepler* data has so far provided clear answers to the occurrence of small planets (with a radius up to one-half as large as Earth's) around low-mass, cool stars (one-tenth to one-half as massive as our sun—M-type stars). That number is about 0.5 planets per star, or one planet for every two stars considered. When one extends this analysis to stars similar to our sun (K- and G-type stars), one gets an answer of about 0.2 planets per star, or one planet for every five stars considered.

If you like answers to big questions, then they don't come much bigger than that. How many planets orbit a typical star? Well, it might be a number slightly less than one. Billions of stars, billions of planets. This rate of occurrence determined by *Kepler* is likely to be a slightly lower limit on the real rate. Consider how *Kepler* might view our solar system. Although *Kepler* was designed to detect Earth-like planets orbiting a sun-like star, it could not detect a Mars-like planet orbiting the same star (too small). Jupiter would also prove to be undetectable, since its eleven-year orbit about the sun would produce, at most, just one transit within the four-year span of *Kepler*'s observations.

Heaven and Hell

In the foreword to *2001: A Space Odyssey*, Arthur C. Clarke points out a very interesting coincidence: behind every man, woman, and child who walks Earth today stand thirty ghosts, a line of ancestors stretching back to the earliest generations that can be called human. Over one hundred billion souls stalking Earth in search of a home. Clarke points out that the Milky Way galaxy, our home, hosts some few hundred billion stars, and he speculates how many of these distant suns host planets. He considers the possibility that for each human who has ever lived, there exists an individual world each could call his or her own, a private heaven or hell, where each ancestor of humankind could live out eternity. Today's research has revealed as reality Clarke's vision of billions of planets in our Milky Way, their forms astonishing and unexpected.

But among those billions of heavens and hells and all possibilities in between, are there worlds like Earth? Terrestrial worlds with solid rocky surfaces and moderate temperatures, worlds that, depending on the atmosphere present, might support liquid water? Our quest for Earth 2.0, an Earth-mass world, with Earth-like conditions, in orbit about a sun-like star, may well be too narrow, too specific in focus. But what we can say with some certainty from the discoveries of *Kepler* is that Earth-like worlds, terrestrial planets that lie within the habitable zone of their parent stars, are common in our Milky Way. We now know several dozens of such worlds.

Among them are such worlds as Gliese 667 Cc—a worthy character in even the most imaginative science fiction tale. Gliese 667 is a triple-stellar system lying some twenty-three light-years from Earth: stars A and B occupy an inner, binary orbit, and a diminutive companion star, Gliese 667 C—perhaps only 30 percent as massive as our sun—circles them every few hundred years.

This small, faint star is known to host at least two planets, detected by radial velocity variations, and, depending on your faith in faint Doppler signals hidden in the noise, may be accompanied by up to five further planets, for a total of seven. It is Gliese 667 Cc, the second confirmed planet in this system, that attracts our attention. It is at least four times as massive as Earth—most likely a super Earth—and, although orbiting at one-quarter the distance of Mercury from our sun, lies within the habitable zone of its cool parent star.

What a strange and wonderful world. It receives about 90 percent as much stellar energy as we receive on Earth—decidedly clement. Its sun would appear twice as large in the sky as our own—a dull red globe emitting most of its radiation beyond the visible portion of the spectrum to which our eyes have evolved. One habitability problem for planets in orbit about such low-mass, M-type stars is that these stars are intrinsically more active than our sun and regularly disgorge great stellar flares, which could prove hazardous in the extreme for any surface life on a nearby planet.

Too exotic for your tastes? I guess there really is no place like home. What about something more familiar, say Kepler 22b? The first transit from Kepler 22b was discovered only three days into the *Kepler* mission—on May 12, 2009. The parent star, Kepler 22, is located some 620 light-years from Earth and is decidedly sun-like in its properties, being only a shade less massive and cooler than our sun. The radius of Kepler 22b is 2.4 times as large as Earth's, and the planet is likely to be a rocky world.

Kepler 22b orbits slightly closer to its parent star than we orbit the sun. But since Kepler 22 is slightly cooler than our sun, Kepler 22b probably orbits well within the habitable zone: if you compute the equilibrium temperature of the planet, you discover it is 262 Kelvin, or −11° Celsius. Remember, though, that this is the bare-bones temperature of the planet—with no atmosphere to provide a warming greenhouse effect.

What if you assumed that Kepler 22b possessed an atmosphere similar to that of Earth? In this case, the surface temperature would be a pleasant 22° Celsius, slightly warmer than Earth's global average temperature of 15°. That is great—but what if the atmosphere was more like that of Venus, or even Mars? Well, not surprisingly, the additional surface warming provided by these atmospheres would be like that of Venus (way too much) or Mars (not at all), and this is perhaps a good way to appreciate the extremes of actual surface temperature that might exist on Kepler 22b, given our uncertainty regarding what the atmosphere is *really* like.

Despite Kepler 22b being a planet of considerable interest, we unfortunately do not know how massive it is. In addition, though the orbit of the planet is assumed to be circular, there is the possibility that it may be elliptical and thus dips into and out of the habitable zone through its orbit—experiencing potentially dramatic fluctuations in surface temperature as a result.

The Needles and the Haystack

The existence of habitable terrestrial worlds—we can call them Earth-like—provides a powerful human resonance. Although less tangible than the individual worlds I introduced you to above, *Kepler* has also presented us with an equally profound statistical view— one that reveals that such Earth-like worlds populate our Milky Way in their billions. These Earth-like worlds are the reason why exoplanets make it to my top five. Dedicated missions such as *Kepler*, together with patient searches of stellar radial velocities, have revealed specific planetary systems where the conditions to support life might exist.

My language here is deliberately restrained—and "might" is the critical word. In only a few cases can we currently measure the density of a planet and therefore be certain that it is terrestrial as opposed to Jovian. Some of these planets lie within the habitable zones of their parent stars—knowledge that allows us to make at best an approximate calculation of their surface temperatures. An atmosphere and habitable conditions might well exist on one such planet known to us already. Then again, it might not. Even if we discovered 1,000 such potentially habitable, terrestrial worlds there might be 999 stalks of hay and 1 needle. The odds against habitability, let alone life, might be even greater than that.

And I will repeat my warning that any importance we place on searching for life solely on Earth-like worlds is too narrow in its vision. Just think of the worlds in our solar system that are decidedly not Earth-like yet remain compelling to astrobiologists. We find ourselves at the beginning, rather than the end, of our search for life on exoplanets. We have to start somewhere—and terrestrial worlds in the habitable zones of their parent stars are as good a place as any.

We are therefore confronted with a galaxy-sized haystack of terrestrial worlds. Planned space missions to be launched over the next decade will reveal more individual worlds to us. However, beyond the discovery of terrestrial worlds, the next great leap which awaits us is to detect their atmospheres—and to show you exactly why we might want to do this, I have to give you a fleeting glimpse of home.

Galileo: *Can We Detect Life on Earth?*

When the *Galileo* space probe began its journey to Jupiter in 1989, it followed a winding path through the inner solar system. The spacecraft flew by Venus and then Earth, not just once but two times, each gravitational encounter designed to boost the probe's velocity for its long cruise to Jupiter. These close encounters brought with them a unique opportunity. Could this spacecraft, engineered and equipped to investigate the physical environment of Jupiter and its moons, turn its cameras and detectors toward Earth? How would Earth appear to this interplanetary investigator? Could *Galileo*'s instruments detect life on Earth?

Among others, Carl Sagan realized that *Galileo*'s flyby of Earth provided a unique opportunity to mimic an encounter between an interplanetary space probe and a living planet. What would be the result? *Galileo* encountered Earth in December 1990, and three years later a profound paper appeared in the science journal *Nature*. Written by a team led by Carl Sagan, the paper was titled "A Search for Life on Earth from the Galileo Spacecraft." So what did *Galileo* discover?

To begin with, the surface of planet Earth shows a distinctive color, revealed by the onboard spectral imager to absorb strongly through the blue and green portions of visible light. The infrared spectrum, just beyond the visible, is largely unaffected, and the overall signature is referred to as the "red edge." No known rocks or regolith can reproduce this effect—it is the signature of the biological pigment chlorophyll, which evolved over eons to harvest energetic blue-green solar photons for photosynthesis. Infrared photons are less energetic, less useful for photosynthesis, and are simply reflected from the surface of plants to avoid overheating.

The critical observation performed by *Galileo* was to take a spectrum of our atmosphere, revealing abundant molecular oxygen and ozone, together with less abundant—but highly unusual—concentrations of methane. Volcanism, surface chemistry, and photochemistry could not create these levels of oxygen and methane in Earth's atmosphere. Earth's atmosphere appears highly unusual—unusual, that is, if you did not know about Earth's abundant surface life, which processes the atmosphere via the accumulated biochemical reactions that we call metabolism.

Another weird thing—planet Earth chirrups away in the radio portion of the electromagnetic spectrum. Not the short flashes associated with atmospheric lightning, but a strange cacophony of pulsed narrow-band radio signals.[10] Finally, you could ask the question whether *Galileo* obtained any images of the surface that revealed our cities or other man-made features. It turns out that at closest approach, and thus when the smallest details could be resolved, *Galileo* was flying over western Australia and Antarctica and as a result detected no man-made features greater than one kilometer in extent.

These four aspects of the *Galileo* observations—surface color, atmospheric chemistry, radio emission, and surface engineering—have been labeled by some as the Sagan "criteria for life." I think it is pretty clear that any attempt to use all the *Galileo* observations to set precedents for how we search for life beyond ^ arth will place way too much importance on the current state of life on Earth. But the *Galileo* experiment does reveal some broader ideas—ideas that we have encountered before on Mars and Titan: the biochemical processes that define life, and the chemical makeup of any atmosphere in contact with it, evolve as one integrated physical system. This is the key that astrobiologists hope to use to unlock the discovery of life on exoplanets.

Blinded by the Light

So what is stopping us from going out and taking a spectrum of an exoplanet and measuring its atmosphere directly? It turns out that we are not limited by how bright they are. Many exoplanets orbiting naked-eye-visible stars are bright enough that they could be observed with our very best telescopes. Large, Jovian planets would be easier to detect than smaller, terrestrial worlds simply because they reflect more starlight. It is the glare from the parent star—up to ten billion times as strong as the planet's reflected light—that creates the challenge. It swamps the light from the exoplanet. In addition, it is a fundamental limitation of all telescopes and detectors that they cannot create a pinpoint image of a distant star. The star always appears somewhat blurred—to the extent that it washes over the light from the exoplanet, which sinks beneath the glare.

But as a planet transits across the face of its parent star, its atmosphere is briefly backlit. If the atmosphere is transparent, then starlight can pass through it en route to us. As it does so, a certain fraction is absorbed at wavelengths particular to the atoms and molecules of which the atmosphere is composed. If carbon dioxide, water vapor, oxygen, methane, etc., are present, each will leave its absorption fingerprint on starlight passing through. Transits therefore offer momentary yet repeated glimpses through a planet's atmosphere.

So given the best exoplanets discovered to date—those whose parent stars are not too bright and whose orbital periods are not too long (lasting days rather than months)—and given the most sensitive telescopes available, what can we achieve? Can we detect exoplanets' atmospheres? Yes, we can: our ability to capture spectra of exoplanets' atmospheres is keeping pretty impressive pace with our ability to discover ever more Earth-like worlds.

Two particular worlds draw our attention: HAT-P-11b and GJ1214b. Interestingly, these two worlds both sit within the Neptune-to-super-Earth range. HAT-P-11b is twenty-six times as massive as Earth, with a radius four times as large, and GJ1214b is six times as massive as Earth, with a radius that is just under three times as large. Intriguingly, both exoplanets have the same density as the planet Neptune—about one-third that of Earth, or one and a half times the density of water. HAT-P-11b is thought to be a hot Neptune-like world, and GJ1214b is more often presented as a warm, gassy, super Earth—the presence of a large, puffed-up atmosphere making the rocky planet appear larger than it really is (and thus making its density appear lower).

But what can we say of their atmospheres? Each has been observed using the Wide Field Camera 3 on the Hubble Space Telescope, with observations timed to catch each exoplanet as it transits in front of the parent star. The resulting spectra are low-resolution and uncertain—only deep, broad absorption features might be detectable under the best circumstances. Despite these limitations, observations of HAT-P-11b reveal an atmosphere replete with broad water-vapor absorption bands. In contrast, GJ1214b is both intriguing and infuriating in equal measure: the spectrum of the planet's atmosphere is flat and featureless—a result that points to a cloudy atmosphere where the majority of the light from the parent star is reflected rather than transmitted.

Perhaps it is always this way when one attempts to coax new knowledge from nature. Getting these fleeting glimpses of planetary atmospheres offered by transit spectroscopy has taken strenuous efforts, pushing our current telescopes to their limits. Each individual success—such as HAT-P-11b and GJ1214b—represents only fragments of a much larger puzzle: how do atmospheres vary with the mass, temperature, and planetary composition? Hidden somewhere among this puzzle of mixed-up pieces may lie some fleeting clues to life on a single planet. But how to recognize these rare clues? How to find the needle in the haystack?

Lovelock's Dream: Can We Detect Life on Earth 2.0?

To understand whether we can detect life on an exoplanet, we return to an idea of James Lovelock: the atmosphere is a chemical system that may play a role in the biochemistry of alien life. The atmosphere may serve as food, as does the carbon dioxide in our own atmosphere, which photosynthetic planets use to create glucose. It may serve as a waste dump for metabolic garbage—like the oxygen produced by those very same plants or the methane produced by archaea in the digestive systems of ruminant mammals. Whatever biochemical form it takes, surface life—at least on Earth—alters the atmosphere, adds to it, takes from it.

But let me ask you a deliberately tricky question. What is *the* molecule of life? DNA? Perhaps RNA? What about chlorophyll? What about something like oxygen or methane? I can hear the howls of protest already—why focus on one molecule? It is as if I were to ask which chemical element is most associated with life. Carbon? Oxygen? There isn't an element that is unambiguously, uniquely associated with life alone and nothing else.

So let me ask a related question: what molecules should we search for in an exoplanet atmosphere to confirm the presence of life? If you follow your previous, very reasonable protests, the answer would be that there is no single molecule for which one would search. There is no single molecule that provides an unambiguous biomarker.

What about atmospheric oxygen? Haven't I previously advertised the presence of molecular oxygen as a strong biomarker? Let's imagine that our journey of planetary discovery finally reveals Earth 2.0, an Earth-like planet in orbit about a sun-like star. We can go further: imagine that transit spectroscopy reveals the presence of abundant atmospheric oxygen—let's say 20 percent or so. Would this point to the existence of life on this new Earth? I guess that's the key point. It would indeed be a compelling result for astrobiology—but would it be unambiguous? I don't want to be a killjoy, but the answer is no, it would not point unambiguously to the existence of life. True, oxygen is a highly reactive molecule that seeks to bond with almost all other elements it meets. Encountering it in an atmosphere suggests some kind of imbalance whereby oxygen is produced faster than it is consumed. Numerous nonbiological processes can produce molecular oxygen, although in most cases we have encountered, that oxygen reacts quickly with its environment to form new compounds. Though we might be unwilling to consider nonbiological alternatives, the fact remains that in our current state of almost complete ignorance, we have little idea of nature's full repertoire of tricks regarding the role of oxygen in exoplanet atmospheres.

This is the situation in which we find ourselves—and there is no easy answer. Instead, we must understand exoplanet atmospheres in general before we can understand what additional role life may play. What are the effects of non-biological processes: volcanism, photochemistry, surface reactions, etc.? I hesitate to call these processes "ordinary" chemistry, since they are likely to be the dominant processes on most exoplanets that we will study. As we have often found before, the challenge of the astrobiologist is to look again at the chemical properties of exoplanet atmospheres and find the exceptions that do not fit the rule. Which measurements cannot be explained by known physical chemistry? Do the results point to exotic—yet nonbiological reactions—not previously considered? Or might we finally have discovered signs of a living atmosphere?

The Future Is Bright

The good news is that exoplanets are a big deal in both astronomy and astrobiology, and there is no shortage of new and exciting ideas for studying them. Although no single mission will perhaps have the same, astonishing impact of *Kepler*, there is plenty to look forward to.

Among the most ambitious of planned missions is PLATO (Planetary Transits and Oscillations of Stars), the rightful heir to *Kepler*'s throne. This satellite will consist of thirty-two cameras, each observing its own portion of the sky. Whereas *Kepler* observed a single field of view covering 105 square degrees, PLATO will observe two fields covering a total of 4,500 square degrees, or just over 10 percent of the sky. The primary mission goal is to detect up to twenty Earth-like planets orbiting sun-like stars via the same transit method used by *Kepler*. While this may not sound like an astronomically large number, remember that we currently know of no such truly Earth-like planets orbiting sun-like stars. In addition to these prized Earth-like exoplanets, PLATO will detect overall some forty times as many planets as *Kepler* did. That is over one hundred thousand planets—and wins the award for most deserved exclamation mark in this book! Most important, PLATO is a funded ESA mission—to the tune of half a billion euros. Launch is planned for no later than 2024, and the mission lifetime is six years. Although you will have to wait until 2030 to fully appreciate what will be an astonishing exoplanet landscape, I for one am convinced that the wait will be worth it.

Can't wait that long? May I interest you in TESS in the meantime? TESS is NASA's Transiting Exoplanet Survey Satellite, an Explorer-class mission scheduled for launch in 2017. During its two-year mission, TESS will perform the impressive task of searching for planetary transits in over five hundred thousand of the brightest stars across the *whole* sky. Rather obviously, it will not do this in one go, but will instead spend twenty-seven days on each patch of sky viewed by its four-camera array. TESS is expected to detect up to three thousand exoplanets—ranging in size down to Earth-sized worlds around low-mass, M-type stars. Although this may seem to be a less significant number of planets than PLATO's expected one hundred thousand, TESS will characterize *all* short-period transiting planets in orbit around *every* bright star in the sky. And remember, these nearby bright stars and their planets are exactly the kind of targets we require for exoplanet spectroscopy. Once again, TESS is a funded mission—this time a $200 million endeavor—and should more than satisfy your immediate appetite for new exoplanet discoveries.

There is a frustratingly large amount of truly amazing exoplanet ideas that I don't have space to describe. We have taken the first steps toward constructing ground-based telescopes of thirty meters in diameter. Although space telescopes provide stable, continuous viewing conditions, there is a limit to the largest space telescope we can currently fit on top of a rocket. Ground-based telescopes can grow very large indeed—the current giants weigh in with ten-meter mirrors—and provide unequaled light-gathering power. This is a key factor in pushing transit spectroscopy to its limit, and a thirty-meter ground-based telescope will in some aspects outperform even the James Webb Space Telescope, NASA's 6.5-meter successor to Hubble.

What about finally tackling the glare of each exoplanet's parent star? Perhaps the most daring of all exoplanet missions is to fly a giant star shade in space. With the silhouette of a giant sunflower some tens of meters across, the star shade would fly 140,000 kilometers in front of a space telescope such as the James Webb. Much as on a bright day when you place your hand in front of the sun to improve your view of distant objects, this delicate, petal-like structure would act to block all light, even that scattered around the edges of the shade, from the parent star. The artificial eclipse thus created would allow exoplanets to emerge from the glare of their parent stars and reveal themselves to our direct view.

We began this chapter by stepping outside to wonder at the stars in the night sky and considering whether they host new planetary worlds of their own. Our small steps of imagination have been richly rewarded by the confirmation that we exist as one among countless billions of planetary systems.

As astrobiologists, we have asked whether such worlds might host life, and we have debated how to approach this enormous question. Even the closest planets lie at vast distances from us, each lost in the blinding glare of its parent star. Yet we have glimpsed our first evidence of atmospheres in both terrestrial and Neptunian worlds. Though there remains much, much to be learned, we have realized that atmospheric spectroscopy holds the key to unlocking the secrets of exoplanetary life. The opening lines of this particular story have only just been written, and I await further excitement and mystery in equal measure.

The discovery of exoplanets allows us to imagine life spread across the stars of the night sky—separate stellar islands of existence. Would each represent a particular solution to the question, what is life? Or would we instead discover similar features in the separate strands of life, each woven from common threads?

Some astronomers and astrobiologists have dared to speculate further, asking whether intelligent life—of which we hope humans are an example—might also be common throughout the stars. The search for extraterrestrial intelligence takes us into uncertain and at times controversial territory—yet it remains within the realm of science.

nine THE SEARCH FOR EXTRATERRESTRIAL INTELLIGENCE

Ah, SETI. We meet again at last. No other subject within astrobiology has succeeded so well in polarizing the opinions of scientists and those who fund them. Yet arguably, no other endeavor within astrobiology has so captured the public imagination.

Hiding between the familiar radio stations we listen to on Earth lies a world of crackling static. This fizzing hubbub of noise is the sound of the summed radio physics of planet Earth, the sun and planets, and the universe beyond. But what if hidden within the ebb and flow of these radio waves there came an artificial signal—as simple as beep, beep, beep?

The profound realization that occurred to a young generation of radio astronomers growing up in the great age of technology that followed the Second World War was that we had developed the ability to both broadcast and receive radio signals over distances that dwarfed the scale of our galaxy. Yet if we, a relatively young technological civilization, had developed the ability to communicate over such vast distances, could there exist other communicating civilizations beyond Earth? Might the interstellar airwaves be alight with chatter—and might we learn how to listen?

Our small steps that have sought basic life beyond Earth have been overtaken in one giant leap by the effort to discover communicating alien civilizations. Throughout this book, I have argued that our first encounter with alien life is likely to be a meeting of microbes rather than a meeting of minds. We have focused on fundamental tests of metabolic activity or biochemical structure to detect life at its most basic. We now turn to loftier, more technological goals—the detection of life at its most advanced.

To See for a Thousand Light-Years

In 1959, a quite extraordinary article appeared in the journal *Nature*. Sandwiched in between papers on the swarming of bees and the effects of radiation on red blood cells came an unexpected title: "Searching for Interstellar Communications." In just two double-column pages, Giuseppe Cocconi and Philip Morrison outlined the framework followed by almost every SETI project conducted over the next fifty years.

Cocconi and Morrison realized that radio technology offered a practical method of broadcasting signals between stars—and thus a powerful tool with which to recognize the existence of technologically advanced, and therefore intelligent, alien life. A radio telescope can transmit signals as well as receive them. Were we on Earth to transmit a powerful signal into space, that electronic message could be received and recognized by an alien telescope of similar capability at distances up to tens or even hundreds of light-years. Since radio waves are simply long-wavelength photons, they travel at the speed of light. So while any such communication would not be snappy, it might be two-way even over the timescale of a human lifetime.

But even the radio portion of the electromagnetic spectrum offers an unthinkable range of frequencies to be searched. Couldn't we narrow it down somehow? Cocconi and Morrison focused their attention on a frequency of 1420 megahertz. This frequency is closely associated with hydrogen, the most abundant chemical element in the universe. Atoms of hydrogen pervade our galaxy and, to a lesser or greater degree, every other galaxy we have ever observed. A subtle change in the structure of hydrogen atoms, a shift of their excitation state, can release a feeble photon of light. The frequency of the emitted photon is 1420 MHz, and astronomers also know it by its wavelength as hydrogen 21-centimeter emission. Rather like audience members shifting in their seats during a play, the atoms of hydrogen gas present in galaxies are continually shifting their excitation states as a result of interatomic collisions and thereby emitting a soft radio glow of 21 cm radiation.

Hydrogen appeared to offer a pervasive spectral lighthouse that would be familiar to radio observers of the night sky, whether human or alien. Moreover, 1420 MHz is located within an uncluttered portion of the electromagnetic spectrum, where the background radio hiss of our galaxy is relatively quiet. Later astronomers labeled this spectral region the cosmic watering hole: The 21 cm beacon, isolated within a clear patch of the radio sky, seemed to offer the perfect meeting place where civilizations strewn across the galactic wilderness could gather and exchange ideas.

Just like Us?

Radio searches, of individual stars and open swathes of sky, at frequencies close to 1420 MHz have become the familiar approach of almost all SETI projects conducted over the last fifty years. Unfortunately, there is also a rather obvious problem that has dogged all SETI searches since their inception. The arguments presented by Cocconi and Morrison could equally be stated differently: the aliens would use radio telescopes because we would, the aliens would broadcast close to 1420 MHz because we would, the aliens would broadcast to sun-like stars because we would.

We seem to know an awful lot about how the aliens would go about communicating their presence to the universe. We have come across these arguments before—they are called anthropocentric, and they assume that humans occupy some special place in the universe. Anthropocentrism has taken a beating over the last few centuries. We no longer exist at the center of the solar system, the galaxy, or the universe. So why would our ideas concerning interstellar communication be special or unique in any way? The short answer is that they wouldn't be—we just like to think they would. Although I do not anticipate a SETI contact any time soon, my sense—indeed, my hope—is that, when we do receive a message, anthropocentrism will receive yet another shock.

Despite these reservations, SETI excites a powerful resonance within our imaginations. Perhaps a fairer conclusion might be to state that we search because we can, because it is interesting to do so. I have previously drawn attention to the fact that all our current searches for life in the universe—complex or basic—are limited compared with the unthought-of possibilities that might exist. Cocconi and Morrison were among perhaps the first astronomers and astrobiologists to realize that you have to start with what you've got.

Quixotic and anthropocentric as SETI may be, are its foundations any less secure, say, than those of a search that aims to detect basic life existing on so-called habitable exoplanets? That seems to be a pretty pertinent question—especially if those two projects are competing for resources. In time-honored fashion, I intend to answer it with another question: given fifty years of searches for extraterrestrial intelligence, has anyone ever considered what our chance of success might be?

The Drake Equation

In 1961, only two years after Cocconi and Morrison's landmark paper, a small conference was held at the Green Bank Telescope in West Virginia. Its subject was SETI, and just ten people were in the room, drawn from fields as diverse as neuroscience, chemistry, and astronomy. Among them were Philip Morrison and a young Carl Sagan. Together, these pioneers probably discussed many of the ideas we have covered in this book.

The astronomer Frank Drake was present. He had just completed Project Ozma, the first attempt to detect alien radio broadcasts—in this case from the star systems Tau Ceti and Epsilon Eridanus—by using the Green Bank telescope. It was with the intention of summarizing the main agenda item at the meeting that day—what knowledge is required to determine the number of communicating aliens in the Milky Way—that Frank wrote a deceptively simple looking equation on the blackboard.

$$N = R_* \times f_p \times n_e \times f_l \times f_i \times f_c \times L$$

The idea conveyed by the equation can be explained in one rather long sentence: the number of communicating alien civilizations present in the galaxy today (N) is the product of the rate at which new stars are born (R_*), the fraction of them that host planets (f_p), the number of planets per star that are Earth-like (or simply habitable, n_e), the fraction of these planets that develop life (f_l), intelligence (f_i), and the technology to communicate (f_c), and, finally, the lifetime of such civilizations (L).

As I said, deceptively simple, but also deceptively troubling. As one starts from the left-hand side, one can put in numbers that are relatively well known. Take, for example, the rate at which stars are born in the Milky Way—it is approximately four sun-like stars per year. One can refine that number a little bit by considering that most newborn stars are less massive than the sun, but when I talk about a number being relatively well known, I mean, for example, that we are pretty sure it is neither 100 times as large as the value given nor only 0.01 times as large.

That is the good news. As one proceeds from left to right, our knowledge regarding each term in the equation diminishes rapidly. In fact, when considering factors such as the lifetime of a communicating civilization, the best one can come up with is that it is longer than eighty years (the length of time we have been at it) and probably shorter than the age of the universe. Within those limits, your guess is as good as mine. Given such great unknowns, the Drake equation has come in for considerable criticism because any attempt to calculate the number of communicating civilizations is hopelessly swamped by speculation.

Proponents of SETI and the Drake equation often state that Frank Drake originally wrote the equation solely as a means of summarizing the conference agenda that day in 1961. In this sense, one can think of it as a scientific wish list—topics that merit further study if the prospects for detecting communicating aliens are to be quantified. Critics of the Drake equation point to the great unknowns present in the equation and go as far as to question whether their numerical values might ever be known.

THE SEARCH FOR EXTRATERRESTRIAL INTELLIGENCE

So who is right? Is the approach taken by SETI fundamentally flawed—or even worse, unscientific? The simple answer is no. The Drake equation is an effective—all too effective, as it turns out—scientific statement that reveals our considerable ignorance. Moreover, the Drake equation *can and should be* used as a powerful means to quantify the factors that govern the number of communicating civilizations in the galaxy. One should not dodge this problem with evasive arguments. If we are serious about detecting such civilizations, we have to be completely honest about our ignorance and then set about accumulating knowledge to fill the gaps. Yes, some of the factors are, at present, completely unknown, but that does not mean that they are either unimportant or unknowable.

Take, for example, the success of the *Kepler* mission. One way of placing the exoplanet discoveries made by *Kepler* into a broader context is to realize that within twenty years of discovering the first exoplanet, we have relatively secure knowledge of f_p, the fraction of stars that host planets. I am impressed that it took *only* twenty years to achieve this. Big questions take time and effort to solve—and the Drake equation is a highly effective means of breaking down one big question into more easily digestible chunks.

Which do you think is the most important factor in the Drake equation? I would argue that it is always the next number on the right, the one just beyond the limit of our present knowledge. Given what we currently know of the Drake variables, that next unknown is the number of habitable planets per star—a number for which *Kepler* has provided an initial measurement, which both TESS and PLATO will in turn improve on in the future.

One could take a further step to the right and consider the fraction of habitable planets that go on to develop life. This is a big one—perhaps *the* big one—because in order to take any further steps through the Drake equation, we have to discover life beyond Earth. This question confronts all astrobiologists, whether they are searching for bio-sludge on Titan or waiting for a tweet from Tau Ceti. Their chances of success or failure can be represented by the same Drake equation—albeit one more or less truncated compared to the original version.

Personally, I would be (more than) happy to understand the terms of the Drake equation leading to even basic life, let alone quantifying the potential number of new Twitter followers among the stars. Yet no one criticizes other branches of astrobiology on the grounds that the fraction of habitable worlds that go on to develop life is a complete unknown. In fact, one could admit that our efforts to discover basic life beyond Earth are so exciting *because* the question is so big and unknown.

True, the Drake equation as applied to SETI goes much further and requires us to make additional, potentially highly anthropocentric assumptions. Yet when Frank Drake wrote down his equation, he was following good scientific practice. It would be ungrateful of us to complain that in doing so, he revealed a set of profound questions whose answers have so far almost totally eluded us. The agenda he set in 1961 remains just as relevant today and expresses, in powerful simplicity, the full scope of astrobiology.

Serendipity

So what does a SETI search look like? How does one "listen" for signals from distant aliens? One of the most ambitious current SETI experiments is SERENDIP V—the fifth version of the Search for Extraterrestrial Radio Emissions from Nearby Developed Intelligent Populations. The program is coordinated by astronomers at UC-Berkeley and uses a novel configuration of the giant, 305-meter-diameter radio telescope at Arecibo in Puerto Rico.

Arecibo is a wonder of the astronomical world: its spherical reflecting surface is not supported on a steerable dish, but is instead constructed within a natural depression in the local hills. As with all radio telescopes, the reflecting surface is not a mirror in the sense that you or I are used to, but is made of a grid of aluminum sheets that—as viewed by incoming radio photons—forms a continuous reflecting surface.

How do you point the telescope at a particular sky location? The receiver is suspended at the focus of the telescope, 150 meters above the reflecting surface, and is mounted on a nine-hundred-ton platform. By shifting the position of the platform above the reflecting surface, the receiver obtains a slightly different view of the sky. Earth's rotation takes care of the rest, and as a result, Arecibo can view approximately one-quarter of the sky.

The SERENDIP experiment observes by "piggybacking" on the telescope. Operating in parallel with the main receiver, the SERENDIP detector is able to siphon off data without interfering with the primary scientific observations performed at the observatory. In this way, SERENDIP obtains a near-continuous view of the sky, scanning a 200 MHz band of frequencies centered on 1420 MHz.

SERENDIP V collects gigabytes of data every second. Searching for intelligent signals in this ceaseless flow of data is a daunting task—akin to separating wheat from chaff amid an unending bumper harvest. Much of the data processing for SERENDIP V occurs within the instrument's electronics—only signals of potential interest are retained for detailed analysis. The gargantuan computing resources required to cope with this vast amount of data led an innovative team of SETI astronomers in an exciting new direction in 1999.

I imagine that most people may not have heard of SERENDIP, but I am pretty sure that you know about SETI@Home. SETI@ Home takes a portion of the SERENDIP data—a narrow, 2.5 MHz band of frequency space centered on 1420 MHz—and distributes it in bite-sized chunks to public computers whose owners have signed them up for the endeavor. On members' computers, the interaction is managed by a small software package that operates as a screen saver on your computer. You get to enjoying watching the bits crunching away during your idle moments and wait for the big red flashing light to go off.[1] SETI@Home has attracted more than eight million participants since it was launched in 1999 and continues to function today as one of the most powerful supercomputers in existence.

To date, there have been no confirmed signals of extraterrestrial origin from any SETI search. But as in every astronomical search, we have to consider what we could have seen. How sensitive are our observations? The simplest approach is to assume that the aliens operate a duplicate of our equipment—a 300-meter-diameter radio dish capable of sending a signal of total power equal to 1 megawatt. Thought of in this way, the sensitivity of our current SETI searches can be expressed as the distance over which we could detect such a signal. For SERENDIP V, it is approximately one hundred light-years, and for the subset of the data analyzed by SETI@Home, the distance is about three times as far.

This volume of space about planet Earth contains some fifteen thousand stars, most of which will be less massive than our sun (G-, K-, and M-type stars). That is certainly a large number—but it is no longer *astronomically* large. If it turns out that the number of broadcasting aliens is *astronomically* small, you could be a long, long way from ever making a detection. Of course, if the aliens have more powerful equipment, we could detect them over even greater distances. Although, as with many aspects of SETI, you just never know.

The importance of SETI@Home goes far beyond the day-to-day practicalities of SETI searches. By engaging the public directly, the SETI team tapped into much more than mere computing power. As I have said before, SETI is a scientifically well-founded exercise, and any scientific project that involves more than eight million people can only be a good thing. SETI@Home users, I salute you!

Pulling the Plug

Now for a delicate issue. Imagine if, as a scientist, I came to you and told you that I had an instrument capable of measuring an important new physical effect and asked, please, for some money to undertake an experiment. How would you answer? Did I mention that there is a catch? I don't know how likely it is that I will get a result. It might be straightaway. It might be never. I suppose, if it wasn't too expensive, you might say okay, try it once and let's see.

So we do that—and find nothing. I then come back to you and ask for more money to try again. I might claim to have a better detector, but I still can't tell you how likely I am to succeed in making a measurement. Would you keep on giving? Or at some point would you decide that we are getting nowhere fast?

You can see where I am going with this. Let me play devil's advocate for a moment and consider whether one could apply the same arguments, not to SETI, but to the *Viking* missions to Mars. What did we learn from the experiments that made up the biology package? That at the two locations tested there are no life-forms that behave biochemically as the experiment designers expected them to.

How did NASA respond? Did it request funds for two more *Viking* landers, each equipped with substantially the same equipment, and dispatch them to similar locations on the planetary surface? No, it recognized that the question of life on Mars was too big to answer in one go. Instead, NASA focused on the next unknowns on its list: What determines the soil chemistry of Mars? What is the past history of liquid water on the planet? Is there any present today? NASA adapted its approach, asked different questions, and, as a direct result, began to provide valuable knowledge about Mars and its potential as a habitat.

What do we learn about broadcasting aliens if we continue to hear nothing? Perhaps the most penetrating criticism of SETI is that we will not learn anything by conducting further searches for advanced life until we either gain new knowledge of the intervening unknowns along the Drake equation or change our experimental approach. Spending science dollars, finding nothing, and yet learning something can indeed be valuable. Yet spending funds, finding nothing, and learning nothing is a no-no. Public funds via NASA's SETI program were canceled in 1995. Since that time, SETI searches have struggled on through the funding wilderness, supported by private and corporate donations. Surviving but never thriving.

SETI Phone Home

If SETI is your own top astrobiology project, fear not, for wonderful surprises do occur. On July 20, 2015, the scientific philanthropist Yuri Milner announced $100 million in funding for the Breakthrough Listen project. Sharing his platform with such scientific luminaries as Stephen Hawking, Martin Rees (Britain's astronomer royal), Geoff Marcy (a leading planet hunter), and, of course, Frank Drake, Milner appeared to have made SETI's dreams come true: observing time would be contracted from the Green Bank and Parkes radio telescopes to conduct a SETI search some five hundred times as comprehensive as those attempted before.

Will they be successful? I hope you can now appreciate that my comment "your guess is as good as mine" was not an attempt to be facetious but rather expresses our current scientific ignorance on the prospects for intelligent, communicating aliens. Should you already be a member of SETI@Home, you might want to consider a few computer upgrades in order to prepare for an avalanche of new data to analyze. In addition, SETI@Home is now available as SETI@Phone—with apps that can analyze data while your cell phone is recharging at night. One can only hope that millions more citizen scientists will join the ranks of SETI.

Given the reservations and criticisms discussed in this chapter, you might wonder why SETI makes it into my top five scenarios for discovering alien life. Well, when I go to bed each evening, a tiny part of my brain wonders, what if? What if tomorrow I wake up and click on the BBC News website and read the full-page headline that contact has been made? Would I be surprised? Er, yes. Yet would such news be implausible? Clearly not. To paraphrase Cocconi and Morrison, just because I think the chance of success is low, I do not think the basic idea of SETI is flawed.

Would I spend any of my four billion imaginary dollars funding SETI projects? No—not until we make it further to the right in the Drake equation. Am I happy for others to fund it from private sources? Very much so—all power to them.

I began this book by asking why one might choose to read a book about the search for alien life when no such life has been detected. Now that we are nearing the end, you are in a better position to answer this question than I am. I know why I wrote the book: because I find myself fascinated by each small increase in our knowledge, perhaps, if I am honest, even more so than by the imagined moment when new life is discovered. I experience the thrill of each new discovery, even when, at some turns, we discover that we have been going about our search the wrong way. Slowly, tile by tile, we are assembling the pieces of a great puzzle—perhaps the greatest we will ever attempt.

In many ways, this final chapter focuses as much on ourselves as it does on the search for life elsewhere. That's a very egocentric perspective—but a necessary and unavoidable one. It is perhaps impossible to retain a cool sense of detachment as we search for life beyond Earth. If we succeed, we will ultimately gain a unique view of our place in the cosmos. In much the same way, our first journeys to the moon were notable not only as a step in our outward exploration of the solar system but also for the unique and novel view we obtained by looking back at our planet.

What perspective have we gained so far? What questions do we carry with us? What will the answers mean to us? As astrobiologists, can we be more specific? Where are we heading next year? How about next decade? We should each decide—me included—how to spend our imaginary $4 billion budgets.

Though it is both interesting and fun to speculate on future directions in astrobiology, my earnest hope is that there are plenty of surprises waiting for us in the near future. What they may be, I cannot say. But surprises in science are always to be welcomed. They compel us to think in new ways, to view the world around us from a different perspective. Perhaps this is where we should begin, with the idea of perspective.

The Pale Blue Dot

In addition to all its other achievements, *Voyager 1* has provided us with our most distant view of planet Earth. In 1990, as part of a family portrait of images of each of the planets in our solar system, *Voyager 1* snapped a fleeting image of Earth at a distance of approximately six billion kilometers, or 40 times the distance of Earth itself from the sun. Within the image, Earth occupied a single blue-green pixel—a mote of planetary dust suspended in a shaft of scattered sunlight. This is the famous "Pale Blue Dot" that led Carl Sagan to muse on both the folly and the promise of mankind.

Let me confront you with a different perspective—not one based on a new location, but instead one granted by new knowledge. Life is discovered beyond Earth. It doesn't matter which scenario we imagine: Mars, Europa, Titan. An exoplanet, perhaps, or a cryptic yet unmistakable message from an overlooked and unremarkable star system. How would you react? That is a question one can contemplate on a number of levels. First, the practical. Perhaps you read or listen to the news before breakfast. How would the rest of your day differ from the norm? Would you finish your breakfast, or would you stagger outside to gaze at the skies? Would you head to work or instead dash to the supermarket to stock up on bottled water? Would you do the laundry, pay the mortgage or would you reckon that all bets are off?

Depending on the level of contact—and I'm assuming here that they are not hovering directly overhead—would anything change in your daily routine? Your guess is as good as mine. I suspect that the answer is that we would go about our business (almost) as usual. For my own part, my microbreaks spent reading the news might well be (much) longer than usual that day, but life—on Earth—would still go on.

This reaction is not difficult to understand. Think of the scenario whereby life is detected in samples returned from Mars. The knowledge that there exists basic life on Mars, some of which is currently on an extended vacation in an Earth laboratory, will not change the need to respond to the challenges of daily life. Bills will still need to be paid; children will still demand their breakfasts.[1]

Perhaps we need to look beyond the merely practical. Would life be different in a more abstract sense? I think the answer is clearly yes, but exactly how is more difficult to say. Our world would somehow seem larger. I have to admit that I still get excited when I look at the planet Mars and realize that we have two robotic rovers currently exploring the surface. The mind's eye has a far-reaching vision. The prospect of imagining newly discovered life on a distant planet or moon would be even more exciting. But such knowledge would not mark the end of our journey. It would be a step, one just as fundamental as that taken by Neil Armstrong to the surface of the moon. Yet one hopes it would be only one of many in our journey outward from Earth.

Might our new perspective on life change the way we act? Would the small yet fundamental step of discovering life beyond Earth usher in a new stage on our long and ongoing journey to maturity? Perhaps more specifically, if we discover lower life-forms on another planet in our solar system, how will we treat them? What if they are edible? What if their planet is packed with mineral resources much needed on a depleted Earth?

My speculation on this point is likely no more insightful than your own. Throughout this book, we have used life on Earth as a reference point, our only point, in fact, from which to attempt to understand some of the properties of life in general. We can continue using this approach to speculate on how we might interact with new life-forms with which we come into physical contact—and unfortunately, the answers are not very uplifting. One need only consider the fate of edible life-forms on Earth, or those whose habitats provide resources desired by humans. A pessimist (or a realist) might go further and point out the historical fate of human communities of Earth that had the poor fortune to be contacted by those that were technologically superior and expansionist.

I want to avoid an overtly moral tone, but viewed in this way, our past behavior does not point to a happy future. Yet we are bound by no fate, and slowly, incrementally, humanity has evolved ideas and norms—what we call culture—that have advanced our understanding of both nature and our place within it. We have matured. Is there any cause for optimism? Perhaps only the realization that in order to overcome the challenges that currently face the human residents of planet Earth—hunger, overpopulation, warfare, climate change—we may be required to mature further, to develop increased respect for one another and our planet. If this happens, we may take some of that newfound awareness and humility with us beyond Earth in our dealings with new forms of life.

Fundamental Questions

On August 7, 1996, President Clinton stepped onto the South Lawn of the White House to address the assembled press. The occasion was the claim by NASA scientists to have discovered evidence of ancient Martian life in the meteorite ALH84001. His comments at that time are worth recalling here: "It speaks of the possibility of life. If this discovery is confirmed, it will surely be one of the most stunning insights into our universe that science has ever uncovered. Its implications are as far-reaching and awe-inspiring as can be imagined. Even as it promises answers to some of our oldest questions, it poses still others even more fundamental."

These are the words of a consummate politician: uplifting, inspiring—and lacking in any kind of concrete detail. What exactly are those fundamental questions? Doesn't anyone ever write them down? Can we ask some of our own instead?

I for one would take great delight in learning the details of an alien approach to life. Some of my questions would therefore be basic and perhaps obvious: How does this new life-form organize itself and process energy? Does it grow and reproduce? What about inheritance and evolution? Does it do stuff for which we don't yet have words? These are what you might call small-scale questions, ones you answer by pressing your microscope right up against the rock sample or petri dish.

What if we step back and ask some grander questions? What about the following: Is life a natural consequence of complex chemistry, wherever we encounter it? Do we identify life in all environments that offer (say) organic chemicals, energy, and a suitable medium for reactions? What might this tell us about a natural imperative toward life?

Over the last century, our understanding of what life is has progressed through the fields of biology (the study of living systems) to chemistry (the study of the molecules that make up living systems) to physics (the interactions between the atoms that make up the molecules in living systems). Part—but not all—of our current understanding of life has come from viewing life on progressively smaller, one might say more fundamental, physical scales. This is a very reductionist approach to the question "What is life?" one that views life as a series of mathematically expressed relationships between fundamental units of matter—in this case, atoms. But having come so far, let's ask what form these equations might take. What do the laws of "life" physics look like?

To see what I mean by this, consider for a moment the life not of a biological organism but of a star. Formed from a diffuse cloud of gas and the inexorable force of gravity, a protostar collapses inward until the rising temperature and density of the core results in the spark of nuclear fusion. In doing so, the newly formed star goes against the thermodynamic grain of the universe. All matter and energy in the universe is subject to an inexorable process of decay, by which the entropy, or overall state of disorder, must continually increase. We call this rule the second law of thermodynamics.

Throughout their lives, stars exist in a localized state of rebellion against the second law of thermodynamics. Gravity drives stars initially to move toward an ordered state of affairs. Fusion drives the nuclei of atoms to evolve, step by step, along the periodic table—each reaction produces a nucleus more ordered, more complex than its predecessor. The energy liberated by nuclear fusion sustains the star against gravitational collapse. Though it appears that stars break the second law of thermodynamics, in reality "circumvent" might be a more accurate word. Although the star is a highly ordered system, the energy produced by nuclear fusion ultimately leaves the star as a disordered stream of photons. The second law is safe.

Living creatures as we know them are ordered systems of atoms and molecules that also spend their lives in local insurrection against the broader laws of thermodynamics. And much as we saw with stars, disordered energy is returned to the universe as the heat that escapes from our bodies. Both stars and living organisms nonetheless remain exceptionally interesting from the point of view of physics, since each represents a temporary enclave of order formed by the flow of energy.

What might be the forces that lead to the spark of life? Might we come to think of life as a "property" of the universe? Is life a phenomenon that arises from the laws of physics as naturally as stars ignite within dense clouds of prestellar gas? This is clearly a much bigger question, but just as clearly, this is the point at which we should dream big and ask such questions. I don't expect our first, astonishing discovery of new life-forms to provide a definitive answer. Perhaps not even the second, third, or fourth such discoveries. But if we persist in our search, in time answers to questions even as big as these will come.

Keeping Pace with Audacity

Writing a book about astrobiology is both exciting and daunting. Both reactions are due to the sheer volume of new discoveries being made. Even during the year and half I spent writing, there appeared a number of astonishing results. Researchers definitely linked geysers on Enceladus to subsurface reservoirs of liquid water. New statistical techniques developed by the *Kepler* science team resulted in an explosion in the number of known exoplanets. Perhaps we have finally discovered methane on Mars, blown across the surface in enigmatic wafts. The *Rosetta* mission dispatched a lander to a comet and just barely managed to hang on for the ride.

Perhaps the most frustrating aspect of writing was the realization that the book must end at some point while the accumulation of new knowledge will continue. What major discoveries are just around the corner? What current knowledge will soon be out of date? Which space missions will fly? Which will fail to even get off the drawing board? Under what conditions will contact occur? If I claimed to have definitive knowledge, I would be lying. I do believe, however, that the themes outlined in this book—exploration, curiosity, ingenuity, and open-mindedness[2]—will take us a long way toward our goal.

On July 14, 2015, the *New Horizons* space probe flew past Pluto at a distance of ten thousand kilometers. The views have been breathtaking—Pluto has become a world before our eyes. What we thought was a cold, icy relic turned out to be a young, active world powered by as yet unknown reserves of interior heat.

What of the *MAVEN* orbiting mission to Mars? Will it detect the periodic burps of methane sniffed out on the surface of the planet by the *Curiosity* rover? Will it be able to locate their source and provide clues to their origin? Might the site of methane release be within *Curiosity*'s range, or will this quest await a new rover, perhaps one equipped to dig down into the permafrost?

TESS and PLATO will detect new exoplanets—worlds beyond worlds. Each one, akin to a single piece of colored tile, will form part of a mosaic in which we will glimpse the overall pattern of planet formation. Is the mass of the planet the key piece of information that determines its physical nature? Or does the distance at which it forms from its parent star—from which it may well migrate—play a role as well? We will push the technological limits of our instruments to see the faint shadows of planetary atmospheres against the overarching luminosity of their parent stars. Will we discover the long-sought lines of molecular oxygen or ozone? Will this be proof of life, or will it instead serve to open our eyes to the myriad chemical possibilities offered by nonliving planetary atmospheres?

What of our dream of conducting sample return missions to the planets and moons of our solar system? Can we find the resources and motivation to cooperate? Will our enthusiasm survive the fact that our first few attempts, even if they successfully return material, may well have missed any scant traces of biology just by bad luck? Will we skip all this and take one giant leap to visit in person?

Once again, I don't know. The only thing I do know is that we are all going to have to run pretty fast to keep up with all this audacity.

We Could Be Heroes

Back in chapter 4, I gave you a budget of $4 billion to fund an astrobiology program. For each of the five scenarios that I outlined in the subsequent chapters, I tried to give you a sense of how much each endeavor might cost.

While the implementation was playful, the intention was serious. It would be too easy to imagine that we could undertake all the astrobiology projects that we find interesting and deserving. One of the hardest realizations as a scientist (and science funder) is that one cannot do everything. Though it is better to do a small number of projects well rather than attempt many piecemeal, one should likewise recognize that small amounts of money can allow small, innovative ideas to grow into big ones. I therefore wanted to give you a sense of the factors determining which projects get funded and which do not.

Having posed the question to you, I should, out of a sense of fairness, give my own answer. How would I spend my $4 billion? There is no use whining about it being a tough question—that is the whole point! Number one would be a sample return mission to Enceladus. My reasons? Excitement—it's a huge ball of liquid water. Second, it is within our reach—that water is constantly vented into space, and the technology exists to sample it and return it to Earth. Does that mean that I rate the other options as unexciting or unachievable? Certainly not, but as I said, better to do one thing well.

Wait a minute! Do I get any change out of my $4 billion? I might have $1 billion left over—give or take. What could I spend this on to keep smaller projects moving along? I would look to support the technology required to routinely obtain spectra of exoplanet atmospheres. With *Kepler*, TESS, and PLATO, we are well on the way to discovering exoplanets in their thousands—studying their atmospheres is the next open frontier. It may be that an orbiting star shade is the means to achieve this. Alternatively, it may be that building a new generation of overwhelmingly large ground-based telescopes is the way ahead. It is to these projects that I would apply the $1 billion or so left over in my loose-change jar.

Did you wonder why I set the limit at $4 billion? It turns out that in 2013, the nations of the world spent approximately $1.5 trillion on defense. That is $1,500 billion. Defense from what? One another and their own populations. Perhaps the most pertinent observation in Carl Sagan's description of our pale blue dot is of the aggression enacted by the inhabitants of one corner of an Earth-sized pixel upon those who inhabit the other side of the pixel. Viewed in this way, the effort and expenditure used to further such aggression can be seen only as waste on a planetary scale.

Each day an average of $4 billion or more is expended on defense. Every dollar represents a tiny unit of human effort and endeavor. It is a very healthy social exercise to ask to what ends we direct our effort and will—and I am not blindly suggesting that astrobiology as a fundamental science deserves a place at the very front of the queue. But it is profoundly affecting to realize that the aims of this book—or at least your own prioritization of them—could be realized if we asked such questions and were bold enough to seek answers. Just for one day.

NOTES

Chapter 1. Alien Extravaganza

1 Should a future alien be reading this book, please extend me the forbearance of hindsight.

2 Mercury, Venus, Mars, Jupiter, and Saturn.

3 Construction materials, academic and social life, a warm bed.

4 Schiaparelli used an 8.6-inch telescope; Lowell had 12- and 16-inch telescopes at his disposal.

5 But not, of course, for Michael Brown and the International Astronomical Union. Try reading "How I Killed Pluto and Why It Had It Coming" for an excellent history of the decline and fall of a planet.

6 Aha—a new unit! The minimum temperature any physical body can possess is −273° Celsius—where it has no thermal energy whatsoever. This temperature is defined as 0 Kelvin. Since a one-degree change is the same in the Kelvin and Celsius scales, 0° Celsius is 273 Kelvin.

7 What do I mean here? Well, 0.1 differs from 1 by one order of magnitude, or one factor of ten. Take a number such as 1×10^{-10}. It differs from 1 by ten orders of magnitude. Now that is truly a small number when compared with unity.

8 Bacteria and Archaea are two branches in our modern, three-domain view of the tree of life. Bacteria and Archaea are all single celled prokaryotes (cells lacking a distinct nucleus). The third branch consists of the Eukarya and includes all single-celled organisms and larger that contain nucleated cells. You are a eukaryotic organism. Though this three-domain picture of life may seem less intuitive than a classification based on outward features (for example, vertebrae, eyes, opposable thumbs), it forms the simplest and most effective classification of living systems based on their biochemistry.

9 Go ahead—prove me wrong!

10 For example, though there is nothing wrong with dropping an apple upon waking each morning, it is reasonable to expect the law of gravity to still be in force.

Chapter 2. A Universe Fit for Life?

1 Hold out your arm in front of you and look at your little finger. It subtends an angle of approximately 1 degree. The sun (and the full moon, which is much safer to look at) appears half this size in the sky as viewed from Earth.

2 I won't try to reproduce the many good works on Einstein here. Suffice it to say that his 1915 general theory of relativity explained gravity in terms of a reciprocal relationship between matter, energy, and the geometry of space-time.

3 This statement is true for your lifetime, but what of the distant future? Current observations indicate that the expansion rate of the universe is accelerating. If this proves true, then the growth of the cosmological horizon will slow and eventually stop at a very large but essentially fixed value. Every galaxy in the universe not gravitationally bound to the Milky Way will eventually expand beyond this fixed horizon. Assuming that the stars making up our present night sky will have become dark stellar remnants (black holes and cold neutron stars or white dwarfs), the night sky will be dark and desolate beyond consolation.

4 I will be adopting a qualitative approach to the early history of the universe.

5 What if the dinosaur-zapping asteroid had missed? A huge yet interesting speculation that I will not even brush the surface of here.

6 It is also a very good book by Primo Levi.

7 A technical term, in this case representing about one particle in ten million.

8 Strictly speaking, you just need high temperatures to fuse two particles together. But without an accompanying high density of particles, not much fusion will occur.

9 Above some mass, even degenerate neutron pressure will fail. Beyond this point, we believe that a back hole, rather than a neutron star, is formed—the rest of the star still goes supernova.

10 Those heavier than helium.

11 Yes, that means you, Pluto.

Chapter 3. What on Earth Is Life?

1 It is fascinating to note that the mathematician and computer pioneer John von Neumann used a similar definition of (in this case, artificial) life in his 1948 lecture on the general and logical theory of automata— ideas that predate our modern knowledge of molecular genetics.

2 Primarily, the small yet significant presence of carbon dioxide, methane, and water vapor.

3 In addition to a small contribution from the decay of radioactive elements contained within Earth.

4 This is my one-paragraph introduction to evolution. For a deeper understanding of Darwin's theory, I recommend the following two-step approach: read Richard Dawkins' *The Blind Watchmaker*; repeat step 1 until understanding is achieved.

5 For comparison, remember that the oldest meteorites are some 4.54 billion years old, which is considered the upper limit of the possible age of Earth.

6 You may also be familiar with carbon-14 (six protons and eight neutrons). But ^{14}C is radioactively unstable and decays into nitrogen, with a half-life of approximately 5,000 years—effectively instantaneous on geological timescales.

7 The geochemical record does not tell us how this early life converted carbon—generally in the form of CO_2—into fuel. Did it do so using a photosynthetic reaction (either involving oxygen or not) or via a methane-producing reaction (methanogenic)? The data cannot tell. Come back next eon and I'll pick up this thread again.

8 And here it is: $6CO_2 + 6H_2O + energy \rightarrow C_6H_{12}O_6 + 6O_2$.

9 An event that scientific decorum prevents me from calling the great intoxication event.

10 Also known to turn up in nuclear reactors. It is the living embodiment of Nietzsche's statement that what does not kill us makes us stronger.

11 You could, for example, replace Miller and Urey's warm little pond with an experiment designed to re-create the conditions in impact craters or hydrothermal vents. The experimental setup would vary, but the underlying ideas would remain essentially the same.

12 You could devise your own variation of the Miller-Urey experiment— but you would need to reproduce the vacuum of space and exercise considerable patience.

13 A question I have pondered many times while playing the excellent, though thoroughly geeky, board game Primordial Soup.

Chapter 4. A Biological Tour of Our Solar System

1 If only it *were* $64,000 dollars—even my research grant would stretch to that.

2 Imagine a portly Zeus (Jupiter) wearing a tight-fitting T-shirt bearing the word "sun," and you will never again forget the order of the outer planets!

3 The watt is a unit of power, or the amount of energy delivered in a given amount of time. For comparison, a kettle uses approximately two thousand watts of power when boiling your tea. For the aficionado, the total energy used to boil your water is the power multiplied by the time taken to boil. If the power is in watts and the time in seconds, then the energy required is in joules. Units are wonderful things!

4 Also known as the blackbody temperature, following the tradition of the nineteenth-century physicists who originally worked out the math.

5 The section heading describes this theory as reticent mainly because of the giggling reaction of student audiences to any word containing "pan" within it.

6 Do I hear you clamoring for ALH84001—the Martian meteorite claimed to host microfossils of primitive life? Even if I do, you will have to wait for the chapter on Mars before I open that can of microworms.

7 I'm sorry, but I'm going to make you wait until the chapter on Mars for more on that one.

8 The SNC class known as Shergottite, Nakhlite and Chassignite meteorites.

9 *Lunokhod* means "moonwalker."

10 The *Cassini* mission will end in 2017 by mapping Saturn's innermost ring (the D ring) before taking a death dive into the planet's deep atmosphere.

Chapter 5. Mars

1 Upon touching down, one adopts Mars time, with its twenty-two-hour-long day, or sol in NASA speak.

2 My own impressions of the Martian landscape as viewed by *Viking* were more prosaic. On noting the large field of irregular boulders strewn across uneven dips and depressions, I initially thought: you sent an automated lander to perform an unassisted rocket landing on that, and it finished upright and in one piece—you lucky . . .

3 The rover "family tree" is a great example of how multiple missions pursuing a single theme can learn from previous ones and attempt more complex, audacious science than their older siblings—in a way that ambitious single missions cannot.

4 The fine dust of weathered minerals on planets and moons. It is not really soil as we know it on Earth, since such environments lack decayed biological material.

5 For comparison, the average depth of Earth's oceans is 3,600 meters. This would decrease to 2,600 meters if we flattened out Earth's surface to create a global ocean. Only about 3 percent of Earth's water reserves are locked up in the polar ice caps. Melting them would add a proportionally small amount to the average ocean depth—about 100 meters—though potentially with a very significant impact for the 40 percent of Earth's population living within 100 kilometers of a coast.

6 The average atmospheric pressure of Mars allows liquid water to form only over a very narrow range of temperatures. Instead, exposed water ice typically passes directly from the solid into the gas phase.

7 Though not covering the whole planet: Mars is lopsided, the southern hemisphere being on average at a higher elevation than the north. As a result, Mars would have had a large northern ocean with pockets of water elsewhere.

8 On a more technical level, another concern is that the claimed spectral line emission of methane is narrower than the spectral resolution of the instrument used to detect it—another characteristic of a marginal detection.

9 Another way of looking at this is to consider whether Mars, like our hero Westley in *The Princess Bride*, may only be "mostly" dead.

10 Try Googling "Curiosity REMS" to get today's weather on Mars (REMS is the rover environmental monitoring station).

11 Literally "hidden within stone."

12 Technically, what I mean is that these rocks have a lower albedo, or reflectivity. Consequently, they absorb a greater amount of solar energy than high-albedo rocks and therefore achieve warmer daytime temperatures.

13 A very dubious pun at this point in the discussion.

14 Although the process of discovering them is a little more involved. Being darker than the surrounding ice, the meteorites heat up and melt their way into the ice sheet. But the continental flow of the Antarctic ice sheet is interrupted by the Transantarctic Mountains. Local conditions can return the meteorites to the surface and concentrate them in relatively small areas. Keen-eyed scientists from the U.S. ANSMET (Antarctic Search for Meteorites) program scour the fringes of the Transantarctic Mountains on extended snowmobile tours and return with up to 130 kilograms of meteorites in a given season. This was how ALH84001 was discovered.

15 Some might claim that the Martian meteorite dubbed "Black Beauty" is older. While it does contain tiny crystals of zircon with radiometric ages of 4.4 billion years, these are tiny fragments embedded within a relatively youthful 2-billion-year-old breccia.

16 And the trail would eventually lead to Bill Clinton on the White House lawn (I return to that one at the end of the book).

17 Though one cause for concern is that the sample also contains amino acids clearly of terrestrial origin.

18 The whole story is told in an entertaining and informative appendix to Schopf's book *The Cradle of Life*.

19 To put this value in context, escape velocity is 11 kilometers per second on Earth, and 2.4 kilometers per second on the moon.

20 My politeness filter is fixed firmly in place!

21 At least not if you have equipped them with moderately radiation-hardened electronics.

Chapter 6. Europa and Enceladus

1 They had learned from the *Pioneer* probes' experiences of Jupiter's hazardous radiation environment, and were better engineered as a result.

2 What this means is that if you were riding along with the probe, you would weigh 230 times as much as you do on Earth's surface.

3 Though not entirely isolated. The small background magnetic field generated by *Galileo* itself—which could mask the presence of faint signals—was cunningly removed by having the magnetometer boom rotate about the main structure of the craft. This introduced a clear time and orientation variation into the measured field that proved easy to recognize and remove from the data.

4 Callisto is not yet in resonance with the inner moons but will become so in the future as the orbits of the inner moons grow larger and come to influence their outermost sister.

5 Rather fortunately, the windows of the submersible—made of the same plastic—did not suffer the same fate.

6 Or five-meter wavelength if you prefer it that way.

7 What exactly do I mean by environmentally sensitive? It doesn't make sense to go to all that effort to explore a pristine subglacial lake if, in the process of drilling down and obtaining your samples, you introduce surface bacteria into the environment you are trying to investigate. It goes beyond mere conservation and gets to the heart of your scientific conclusions: if you discover a new kind of bacterium living in the lake, how

can you be sure that you didn't bring it there with you? This is an example of forward contamination (discussed later). How can we preserve the intrinsic purity of the alien environments we are currently exploring? The U.S. Antarctic exploration team that drilled into Lake Whillans in 2013 did so with a hot-water drill that used both bacterial filters and sterilizing ultraviolet radiation to reduce the likelihood of possible forward contamination of the lake environment. Their sampling equipment was decontaminated to similarly high standards. The Lake Vostok team used a more conventional drill lubricated with kerosene and Freon. In the view of other international teams, this introduced a considerable risk of forward contamination. The debate still smolders.

8 This statement is not intended to denigrate hitchhikers—of which I am one—just the delinquent ones who ride on space probes.

9 In addition to flags, smallpox, etc.

10 One could always argue that any kind of planetary protection from sample return missions is unnecessary given the many tons of solar system material that rain down on Earth each day in the form of dust and micrometeorites. Of course, almost all such material is reasonably expected to be nonbiological in nature—though we reserve the right to be surprised.

11 Though one might call this name unimaginative, in the words of a popular British floor sealant of the 1990s, "It does exactly what it says on the tin."

Chapter 7. Titan

1 A wonderful quotation from Ovid ("They brought the distant stars closer to our eyes") plus a collection of seemingly random letters—still technically an anagram, but points are deducted for those loose letters strung on the end. Admittedly, the best I could come up with to disguise "Saturn has a moon!" was "Hot Roman Saunas!"

2 Bemused by what to call the abundant brown sticky sludge produced in their attempts to re-create the organic chemistry of Titan and other solar system environments, Sagan and Khare opted for tholins, from the Greek *tholos*, or muddy (the ancient term closest to sticky brown goop).

3 Don't get confused here. The solar wind is made of energetic charged particles—mainly protons—that are deflected from Titan by Saturn's magnetic field. Photons are electrically neutral and do not experience any magnetic force.

4 One ESA scientist likened the experience to landing in crème brûlée—perhaps my favorite image of the book.

5 Interestingly in this case, the fact that *Cassini's* radar bathymetry is able to measure such a profound depth indicates that the lake must be composed of almost pure methane.

6 One wonders exactly where that next mission would be sent.

7 One could call it the Sauron of early, as opposed to Middle, Earth.

8 This is ingeniously achieved by confining the charged gas particles in an electrostatic box—an electrically charged wire cage to you and me. Electric repulsion between the wires and gas particles keeps the gas in place.

9 At a stately walking pace of approximately one to two meters per second.

10 One astronomical unit is defined as the mean Earth-sun distance over the course of a year.

11 Even though the *Voyager* craft are not heading toward Alpha Centauri, it remains instructive to consider their journey relative to our nearest star system.

Chapter 8. Exoplanets

1 If you think about it, such a planetary system would be difficult to detect with the Doppler wobble technique as well. In such a head-on planetary system, the motion of the star would be perpendicular to us. Only the component of the star's motion that is along our line of sight— we call this the radial component—would generate a signal detectable by using the stellar radial velocity technique.

2 They are—deep breath—tetrahedron (four equilateral triangles), cube (six squares), octahedron (eight equilateral triangles), dodecahedron (twelve regular pentagons), and icosahedron (twenty equilateral triangles). Each beloved alike by geometers and by fans of Dungeons and Dragons.

3 In exact terms, Newton demonstrated that $P^2 = [4\pi/G(M_{star} + M_{planet})]a^3$ but since M_{star} is normally very much greater than M_{planet}, you are free to ignore M_{planet}.

4 This is what scientists call a five-sigma limit, in which the chance of such a detection arising from a random blip in the data is 1 in 3.5 million.

5 *Photo-metry*—the measurement of brightness.

6 Perhaps they took heart from Kepler the man. When developing his orbital system based on Platonic solids, he wrote to the Duke of Württemberg for funds to construct a physical model. He extolled its mathematical and physical beauty, suggesting that the resulting structure

could be ornamented with jewels and perhaps double as a chalice. Un-moved, the duke's secretary replied that though it was an engaging idea, they were unsure that the building of such a model was feasible. They suggested that Kepler start with a paper replica—a formula fol-lowed by funding agencies to this day.

7 Stars can be classified into a temperature sequence, each labeled with a single letter, OBAFGKM, from the hottest O stars to the coolest M stars. It turns out that this sequence is also closely related to both lumi-nosity and mass: O-type stars are the most massive and luminous, whereas M-types are among the least luminous and massive of so-called normal, or main sequence, stars.

8 Though Google tells me it is a quincunx of planets. This seems much too Harry Potter for my liking. Though the same website informs me that the collective noun for professors is a pomposity, so there may be merit to this list after all.

9 Transit timing variations, known as TTV, measure the masses of plan-ets in multiple planetary systems by observing how their gravitational interactions affect their orbits. In simple terms, the time at which each planet transits in front of its parent star varies slightly, since each is either tugged forward or pulled backward in its orbit by its planetary siblings.

10 "Livin' on a Prayer," anyone?

Chapter 9. The Search for Extraterrestrial Intelligence

1 Sadly, this is not a feature of the current version of SETI@Home.

Chapter 10. The Meaning of (Alien) Life?

1 I know mine will.

2 Plus bloody-mindedness if you happen to be in the process of propos-ing a new space mission or telescope project.

REFERENCES: BILLIONS AND BILLIONS OF BOOKS

I don't want to provide a list of all the books I read before writing this one. I would rather give you a list of books, journal articles, and websites that you might find interesting. I hope that these suggestions will provide some additional stimulation, insight, or plain, hard facts that will enable you to extend your interest in astrobiology.

Chapter 1

No one book captures so well our place in the universe as *Cosmos* (New York: Random House, 1980) by Carl Sagan. In the same vein, less lyrical but filled with more hard science (and great images) is *Philip's Atlas of the Universe* (London: Philip's, 2006) by Patrick Moore. I have both these books on my shelves, and I constantly find new ideas and insights within them.

Instead of a biography of Percival Lowell or an account of the Mars canal controversy, take a look at the short yet fascinating book *Is Mars Habitable?* (Macmillan, 1907), by Alfred Russel Wallace. As well as being the cocreator of the theory of evolution, Russel (who was eighty-four when this book was published) provides a detailed and durable rebuttal of Lowell's ideas via the application of basic physics to observations of Mars.

Here is the journal reference for the academic paper announcing the discovery of the first extrasolar planet, in this case one orbiting the star 51 Pegasi: M. Mayor and D. Queloz, "A Jupiter-mass companion to a solar-type star," *Nature* 378 (1995): 355–59.

Have you ever read a good science fiction story? If not, try *The War of the Worlds* (Heinemann, 1898) by H. G. Wells. It is a classic astrobiology text!

Chapter 2

Two books provide an interesting and human perspective on the development of modern cosmology. The first is *Lonely Hearts of the Cosmos* (Boston: Back Bay, 1999) by Dennis Overbye, and the second is *The Day without Yesterday* (New York: Basic Books, 2005) by John Farrell. Although the universe existed long before humans, cosmology as a science is a human endeavor, and these books tell that story and teach much good cosmology along the way.

If you want to explore cosmic chronology in more detail, I offer two suggestions, one scientific, the other poetic. *The First Three Minutes* (New York: Basic Books, 1979) by Steven Weinberg offers a tremendously accessible account of the particle and nuclear physics that occurred in the first few minutes of the history of the universe. *The Periodic Table* (New York: Schocken, 1984) by Primo Levi, particularly the chapter "Carbon," tells the story of the universe from the perspective of a single atom of carbon.

The following academic paper tells the scientific story of the creation of the elements by using the theory of stellar nucleosynthesis: E. M. Burbidge, G. R. Burbidge, W. A. Fowler, and F. Hoyle, "Synthesis of the Elements in Stars," *Reviews of Modern Physics* 29 (1957): 547–650.

Chapter 3

Perhaps the best, most easily accessible reference (that is, not a hard-to-find journal article) for Earth's age is *Ancient Earth, Ancient Skies: The Age of the Earth and Its Cosmic Surroundings* (Stanford, Calif.: Stanford University Press, 2001) by G. Brent Dalrymple.

Have you read *The Blind Watchmaker* (New York: Norton, 1986) by Richard Dawkins? If not, do so—more than once if necessary!

The Cradle of Life (Princeton, N.J.: Princeton University Press, 2001) by William Schopf provides a great discussion of how microfossils provide evidence for the existence and properties of ancient life on Earth.

Take a look at the following back-to-back letters to *Nature* (vol. 416, March 7, 2002) for a perspective on the scientific debates occurring at the cutting edge of such research: "Laser-Raman Imagery of Earth's Earliest Fossils" by William Schopf et al. and "Questioning the Evidence for Earth's Earliest Fossils" by Martin Brasier et al.

A comprehensive paper on the use of geochemical carbon isotope ratios to infer the history of life on Earth is provided by M. Schidlowski, "Carbon Isotopes as Biogeochemical Recorders of Life over 3.8 Ga of Earth History: Evolution of a Concept," *Precambrian Research* 106 (2001): 117–34.

An excellent description of research into Earth's early atmosphere is provided by J. Kasting, "Earth's Early Atmosphere," *Science* 259 (1993): 920–26.

The original paper detailing Stanley Miller's research is S. L. Miller, "Production of Amino Acids under Possible Primitive Earth Conditions," *Science* 117 (1953): 528–29 (most easily accessible from the *Wikipedia* page for the Miller-Urey experiment).

For a description of the results of NASA's *Stardust* sample return mission, take a look at D. Brownlee et al., "Comet 81P/Wild 2 under a Microscope," *Science* 314 (2006): 1711–16.

Stephen Benner provides a broad and cutting-edge description of our biochemical understanding of the origin of life in *Life, the Universe, and the Scientific Method* (FfAME, 2008).

Chapter 4

Two papers on anoxygenic photosynthesis in low-light environments are J. Overmann, H. Cypionka, and N. Pfennig, "An Extremely Low-Light-Adapted Phototrophic Sulfur Bacterium from the Black Sea," *Limnology and Oceanography* 37 (1992): 150–55, and S. A. Crowe et al., "Photoferrotrophs Thrive in an Archean Ocean Analogue," *Proceedings of the National Academy of Sciences* 105 (2008): 15938–43.

Results for the exposure of microbial samples on the ISS EXPOSE facility can be found in S. Onofri et al., "Survival of Rock-Colonizing Organisms after 1.5 Years in Outer Space," *Astrobiology* 12 (2012): 508–16.

Gary Flandro's original paper expounding the gravity-assist route to the outer solar system is contained in G. Flandro, "Fast Reconnaissance Missions to the Outer Solar System Utilizing Energy Derived from the Gravitational Field of Jupiter," *Astronautica Acta* 12 (1966): 329–37. The easiest way to obtain a copy is from the references on the *Wikipedia* page for Gary Flandro.

The best way to get up-to-date information on current space missions is to Google them—in most instances, you will be taken to the mission home pages (mainly) at NASA or the ESA. So Google "New Horizons," "Mars Reconnaissance Orbiter," "Mars Curiosity," and "Cassini."

When taking a look at the *Mars Reconnaissance Orbiter* (*MRO*) pages, keep an eye out for "Be a Martian," a citizen science venture that gives you the opportunity to study, classify, and mosaic *MRO* images.

Chapter 5

Lots of results papers here.

Many papers detail discoveries of potentially water-driven features on Mars using orbital observatories. Here is one example describing the discovery of the ancient Eberswalde delta on Mars: M. C. Malin and K. S. Edgett, "Evidence for Persistent Flow and Aqueous Sedimentation on Early Mars," *Science* 302 (2003): 1931–34.

Mars Odyssey reveals near-surface hydrogen-rich soil across the Martian globe: W. C. Feldman et al., "Global Distribution of Near Surface Hydrogen on Mars," *Journal of Geophysical Research* 109 (2004): 2156–202.

Results from *Phoenix:* P. H. Smith et al., "H_2O at the Phoenix Landing Site," *Science* 325 (2009): 58–61.

Here is the academic paper describing the results of the *Viking* labeled release experiment: G. V. Levin and P. A. Straat, "Viking Labeled Release Biology Experiment: Interim Results," *Science* 194 (1976): 1322–29.

Instead of providing you with all of the results papers dealing with methane on Mars, here are two skeptical views from Kevin Zahnle, which sum up much of the observational data: K. Zahnle, R. S. Freedman, and D. C. Catling, "Is There Methane on Mars?," *Icarus* 212 (2011): 493–503, and K. Zahnle, "Play It Again, SAM," *Science*, 347 (2015): 370–71.

Seasonal gullies on Mars: A. S. McEwen et al., "Recurring Slope Linae in Equatorial Regions of Mars," *Nature Geoscience* 7 (2013): 53–58.

The survivability of Earth microbes on Mars is discussed in W. L. Nicholson, K. Krivushin, D. Gilichinsky, and A. C. Schuerger, "Growth of *Carnobacterium* spp. from Permafrost under Low Pressure, Temperature, and Anoxic Atmosphere Has Implications for Earth Microbes on Mars," *Proceedings of the National Academy of Sciences* 110 (2013): 666–71.

Here is the original paper describing the Martian meteorite ALH84001: D. S. McKay et al., "Search for Past Life on Mars: Possible Relic Biogenic Activity in Martian Meteorite ALH 84001," *Science* 273 (1996): 924–30.

The U.S. Planetary Decadal Survey for 2013–2022 can be found on NASA's website on the "Science Strategy" page: https://solarsystem.nasa .gov/2013decadal.

Chapter 6

Linda Morabito was the NASA engineer who discovered the first evidence of active volcanism on Io. Her account of the process that led to the discovery is given in the following article: http://arxiv.org/abs/1211.2554.

The utterly excellent textbook *Planetary Science* (Bristol, UK: IOP, 2002) by George Cole and Michael Woolfson contains a very clear description of the physics giving rise to volcanism on Io.

Here is the journal article announcing the discovery of subsurface oceans on the moons Europa and Callisto: K. K. Khurana et al., "Induced Magnetic Fields as Evidence for Subsurface Oceans in Europa and Callisto," *Nature* 395 (1998): 777–80.

For two articles describing the discovery and biological implications of deep-ocean hydrothermal vents, see R. D. Ballard, "Notes on a Major Oceanographic Find," *Oceanus* 20 (1977): 35–44, and H. W. Jannasch, "Chemosynthetic Production of Biomass: An Idea from a Recent Oceanographic Discovery," *Oceanus* 41 (1998): 59–63. (*Oceanus* is the magazine of the Woods Hole Oceanographic Institute.)

Keep up to date with the Lake Whillans subglacial research project at the website of the Whillans Ice Stream Subglacial Access Research Drilling, www.wissard.org.

The academic paper describing the discovery of active geysers on Enceladus can be found at: C. J. Hansen et al., "Enceladus' Water Vapor Plume," *Science* 311 (2006): 1422–25.

Here is an academic paper describing the concept of a sample return mission to Enceladus: P. Tsou et al., "LIFE: Life Investigation for Enceladus; A Sample Return Mission Concept in Search for Evidence of Life," *Astrobiology* 12 (2012): 730–42.

Chapter 7

The science journal *Nature* has very sensibly provided an online web focus issue devoted to *Huygens* on Titan: www.nature.com/nature/focus/huygens. All the articles are free access.

What exactly is a tholin? Check out C. Sagan and B. N. Khare, "Tholins: Organic Chemistry of Interstellar Grains and Gas," *Nature* 277 (1979): 102–7.

Rare Earth (New York: Copernicus, 2000) by Peter Ward and Donald Brownlee provides a particular view of the possibilities for life beyond Earth.

For the academic paper accompanying Chris McKay's idea of "Rare Titan," see C. P. McKay, and H. D. Smith, "Possibilities for Methanogenic Life in Liquid Methane on the Surface of Titan," *Icarus* 178 (2005): 274–76.

Peter Ward and Steven Benner provide a very useful discussion of possible alien biochemistries in their contribution to *Planets and Life: The Emerging Science of Astrobiology*, edited by Woodruff T. Sullivan and John A. Baross (Cambridge: Cambridge University Press, 2007), 537–44. This article can be accessed from Steven Benner's list of publications on his home page at the website of the Foundation for Applied Molecular Evolution, www.ffame.org.

Miller-Urey processes on Titan? Check out the following academic paper: S. M. Horst et al., "Formation of Amino Acids and Nucleotide Bases in a Titan Atmosphere Simulation Experiment," *Astrobiology* 12 (2012): 1–9.

Here is the paper that considers the flow of hydrogen in Titan's atmosphere: D. Strobel, "Molecular Hydrogen in Titan's Atmosphere: Implications of the Measured Tropospheric and Thermospheric Mole Fractions," *Icarus* 208 (2010): 878–86.

Chapter 8

Having decided to focus on the transit method for detecting exoplanets, I shamefully skip over a great many alternative approaches. A useful resource for remedying this bias on my part is to take a look at the special issue of *Science* (May 3, 2013) that provides a broad, well-informed review of the techniques for discovering exoplanets.

Keep up with the Kepler space mission at kepler.nasa.gov.

Here is the academic paper presenting the discovery of multiple planets in the Upsilon Andromeda system: R. P. Butler et al., "Evidence for Multiple Companions to Upsilon Andromedae," *Astrophysical Journal* 526 (1999): 916–27.

Here is the academic paper describing the Kepler 10 planetary system: X. Dumusque et al., "The Kepler-10 Planetary System Revisited by HARPS-N: A Hot Rocky World and a Solid Neptune-Mass Planet," *Astrophysical Journal* 789 (2014): 154–67.

For more details on the statistical occurrence of planets about stars as informed by the *Kepler* mission, see: N. M. Batalha, "Exploring Exoplanet Populations with NASA's Kepler Mission," *Proceedings of the National Academy of Sciences* 111 (2014): 12647–54.

Habitable exo-Earths? Try the following papers for more information on Gliese 667Cc and Kepler-22b: G. Anglade-Escudé et al., "A Dynamically-Packed Planetary System around GJ 667C with Three Super-Earths in Its Habitable Zone," *Astronomy and Astrophysics* 556 (2013): 126–49, and W. J. Borucki et al., "Kepler-22b: A 2.4 Earth-Radius Planet in the Habitable Zone of a Sun-Like Star," *Astrophysical Journal* 745 (2012): 120–35.

Can we detect life on Earth? See C. Sagan et al., "A Search for Life on Earth from the Galileo Spacecraft," *Nature* 365 (1993): 715–21.

Here are two great papers that present some of the best results we can achieve for the spectroscopy of exoplanet atmospheres: J. Fraine et al., "Water Vapor Absorption in the Clear Atmosphere of a Neptune-Sized Exoplanet," *Nature* 513 (2014): 526–29, and L. Kreidberg et al., "Clouds in the Atmosphere of the Super-Earth Exoplanet GJ1214b," *Nature* 505 (2014): 69–72.

Chapter 9

Here is where it all began: G. Cocconi, and P. Morrison, "Searching for Interstellar Communications," *Nature* 184 (1959): 844–46.

For a contemporary article on Frank Drake's pioneering Project Ozma, look up F. D. Drake, "Project Ozma," *Physics Today* 14 (1961): 40–46.

Here is where it all ended, a discussion of the events that led to the canceling of NASA's SETI funding: S. J. Garber, "Searching for the Good Science: The Cancellation of NASA's SETI Program," *Journal of the British Interplanetary Society* 52 (1999): 3–12.

For more information on current SETI searches, see D. Werthimer et al., "The Berkeley Radio and Optical SETI Program: SETI@home, SERENDIP, and SEVENDIP," in *The Search for Extraterrestrial Intelligence (SETI) in the Optical Spectrum III*, edited by Stuart A. Kingsley and Ragbir Bhathal, Proceedings of SPIE 4273 (2001).

Type "SETI@Home" into Google and take your first steps into a wider world.

Chapter 10

Although Carl Sagan's text on the pale blue dot appears in many places on the Internet, the material is taken from his book *Pale Blue Dot: A Vision of the Human Future in Space* (New York: Random House, 1994).

Here are two very different papers musing on the role played by the second law of thermodynamics in life. The first is older and more general yet still pertinent today: J. Bronowski, "New Concepts in the Evolution of Complexity: Stratified Stability and Unbounded Plans," *Synthese* 21 (1970): 228–46. The second paper is more recent and contains some wonderfully clear math: C. H. Lineweaver and C. A. Egan, "Life, Gravity, and the Second Law of Thermodynamics," *Physics of Life Reviews* 5 (2008): 225–42.

Mars Hand Lens Imager
(MAHLI), 75
Mars Odyssey, 77, 80, 91
Mars Reconnaissance Orbiter
(*MRO*), 65–66, 76, 79, 89, 90
Mars Science Laboratory. See
Curiosity rover
MAVEN (*Mars Atmosphere and
Volatile Evolution*), 84, 186
McKay, Chris, 130, 138
Mercury, 26, 55, 57, 58
metabolism: origin of, 48–49
metals, 26
meteorites, 46–47, 191n5; in
Antarctica, 193n14; from Mars,
67, 92–95, 194n15
methane/methanogens; in Earth's
atmosphere, 39–40, 82; in
Mars's atmosphere, 82–84; in
Titan's atmosphere, 70, 127–28,
129, 132, 133, 138
microbes. *See* archaea; bacteria
Milan observatory, 6
Milky Way galaxy, 27; planetary
systems in, 8–11
Miller, Stanley, 44, 134
Miller-Urey experiment, 44–45,
46–47, 191n11; as applied to
Titan, 133–35
Milner, Yuri, 179
missions. *See* space missions
MOM (*Mars Orbiter Missions*), 84
montgolfière (hot air balloon),
140–41
moon, the: evidence of volcanic
activity on, 102; first observations of, 4–5; human missions
to, 102; the surface of, 36; as a
world, 5

Morrison, Philip, 15, 170, 172
Murchison meteorite, 46

NASA. *See* space missions; *and
names of specific missions*
natural selection, 34
Neptune, 55, 72, 143
Neumann, John von, 190n1
neutrinos, 54
New Horizons space probe, 186
Newton, Isaac, 149, 196n3
nitrogen: on Titan, 127, 133, 134
nuclear fusion, 8, 23, 25, 184–85;
in the core of the sun, 54

oceans: and the Miller-Urey
experiment, 44
Olbers, Heinrich, 17, 20
Oort cloud, 143
Oparin, Alexander, 43–44, 45
Opportunity rover, 68, 76
orbiter space missions, 65–66.
See also *Galileo* space probe;
*Mariner 9; Mars Express Orbiter;
Mars Reconnaissance Orbiter*
(*MRO*); *Trace Gas Orbiter*
oxygen: in Earth's atmosphere,
38–39
ozone, 40

panspermia, 62–63
Pegasus, 9
periodic table, 22, 48
Phanerozoic eon, 35, 41
Phobos (moon of Mars), 55;
mission to, 68–69
Phobos-Grunt mission, 68–69
Phoenix lander, 78–79
photons, 54